4th Biomedical Natural Language Processing Shared Task Workshop (BioNLP-ST 2016)

Held at ACL 2016

Berlin, Germany
13 August 2016

ISBN: 978-1-5108-2765-3

ACL 2016

The 4th BioNLP Shared Task

Proceedings of the 4th BioNLP Shared Task Workshop

August 13, 2016
Berlin, Germany

Introduction

The BioNLP Shared Task series pursues establishing a community-wide effort for fine-grained information extraction (IE) in biology domains. As in the third edition (2013), the fourth edition focuses on knowledge base construction through IE. Nowadays some BioNLP tools, e.g., protein name recognizers or protein-protein interaction extractors, are reaching a level of applicability which allows them to be integrated into bioinformatics systems and thus to significantly contribute to the bioinformatics and biology research. The BioNLP community then faces the challenge of the integration of information extracted from the text to external resources like -omics databases, biological ontologies, or systems biology operational models. BioNLP-ST 2016 offers tasks by which new or existing data, and evaluation methods are expected to be adapted to this trend.

BioNLP-ST 2016 comprises three tasks that address different aspects of knowledge acquisition from text and also encompasses a wide range of biological diversity:

- SeeDev, which aims at extracting the regulation of the seed development in plants using a rich model;

- Bacteria Biotopes 3 (BB3) for the construction of a bacteria habitat database using external ontologies;

- Genia 4 (GE4), which aims at delivering new shared task framework to construct knowledge base of NFκB synthesis and regulation through IE.

The previous editions had attracted many participants and it has then supported active research on information extraction and database/knowledge base integration. In this edition, 26 final results were submitted by 15 distinct teams from 10 different countries of four continents. This year, BioNLP-ST is organized as a joint event with the BioASQ challenge that has converging goals on biological question answering and semantic indexing. The BioNLP-ST/BioASQ workshop is collocated with the BioNLP workshop hosted by the ACL/HLT 2016 conference in Berlin, Germany. In addition to the participating systems, an overview of each task is also presented at the workshop.

Thanks to the many excellent manuscripts received from participants and the efforts of the programme committee, it is our pleasure to present these proceedings that describe the BioNLP Shared Task and the participating systems.

Claire Nédellec, Robert Bossy and Jin-Dong Kim

Organizing Committee:

Claire Nédellec (INRA) - **Chair**
Robert Bossy (INRA)
Jin-Dong Kim (DBCLS)

Program Committee:

Robert Bossy (INRA) - **Chair**
Jin-Dong Kim (DBCLS) - **Chair**
Fabio Rinaldi (University of Zürich)
Jari Björne (University of Turku)
Jörg Hakenberg (Illumina)
Tomoko Ohta (Textimi)
Philippe Bessières (INRA)
Filip Ginter (University of Turku)
Roser Morante (Vrije University of Amsterdam)
David McClosky (Google)
Yuka Tateisi (NBDC)
Claire Nédellec (INRA)
Georgios Paliouras (Demokritos)
Özlem Uzuner (University of Albany)
Berry de Bruijn (National Research Council Canada)
Anastasia Krithara (Demokritos)
Sabine Bergler (Concordia University)
Jung-Jae Kim (Institute for Infocomm Research)
Kevin B. Cohen (University of Colorado)
Pierre Zweigenbaum (CNRS)
Louise Deléger (INRA)

Task Organizers:

SeeDev Task
Robert Bossy (INRA) - **Chair**
Estelle Chaix (INRA)
Claire Nédellec (INRA)

BB3 Task
Robert Bossy (INRA) - **Chair**
Claire Nédellec (INRA)

GE4 Task
Jin-Dong Kim (DBCLS) - **Chair**
Yue Wang (DBCLS)

Table of Contents

Conference Program

Saturday, August 13, 2016

14:00–14:15   **Overview of BioNLP-ST**

Overview of the Regulatory Network of Plant Seed Development (SeeDev) Task at the BioNLP Shared Task 2016.
Estelle Chaix, Bertrand Dubreucq, Abdelhak Fatihi, Dialekti Valsamou, Robert Bossy, Mouhamadou Ba, Louise Deléger, Pierre Zweigenbaum, Philippe Bessières, Loïc Lepiniec and Claire Nédellec

Overview of the Bacteria Biotope Task at BioNLP Shared Task 2016
Louise Deléger, Robert Bossy, Estelle Chaix, Mouhamadou Ba, Arnaud Ferré, Philippe Bessières and Claire Nédellec

Refactoring the Genia Event Extraction Shared Task Toward a General Framework for IE-Driven KB Development
Jin-Dong Kim, Yue Wang, Nicola Colic, Seung Han Beak, Yong Hwan Kim and Min Song

14:15–15:30   **BioNLP-ST participant session 1**

14:15–14:30   *LitWay, Discriminative Extraction for Different Bio-Events*
Chen Li, Zhiqiang Rao and Xiangrong Zhang

14:30–14:45   *VERSE: Event and Relation Extraction in the BioNLP 2016 Shared Task*
Jake Lever and Steven JM Jones

14:45–15:00   *A dictionary- and rule-based system for identification of bacteria and habitats in text*
Helen V Cook, Evangelos Pafilis and Lars Juhl Jensen

15:00–15:15   *Ontology-Based Categorization of Bacteria and Habitat Entities using Information Retrieval Techniques*
Mert Tiftikci, Hakan Şahin, Berfu Büyüköz, Alper Yayıkçı and Arzucan Özgür

15:15–15:30   *Identification of Mentions and Relations between Bacteria and Biotope from PubMed Abstracts*
Cyril Grouin

15:30–16:00   **Coffee break**

16:00–17:00   **BioNLP-ST participant session 2**

16:00–16:15   *Deep Learning with Minimal Training Data: TurkuNLP Entry in the BioNLP Shared Task 2016*
Farrokh Mehryary, Jari Björne, Sampo Pyysalo, Tapio Salakoski and Filip Ginter

16:15–16:30   *SeeDev Binary Event Extraction using SVMs and a Rich Feature Set*
Nagesh C. Panyam, Gitansh Khirbat, Karin Verspoor, Trevor Cohn and Kotagiri Ramamohanarao

16:30–16:45   *Extraction of Regulatory Events using Kernel-based Classifiers and Distant Supervision*
Andre Lamurias, Miguel J. Rodrigues, Luka A. Clarke and Francisco M. Couto

16:45–17:00   *DUTIR in BioNLP-ST 2016: Utilizing Convolutional Network and Distributed Representation to Extract Complicate Relations*
Honglei Li, Jianhai Zhang, Jian Wang, Hongfei Lin and Zhihao Yang

17:00–17:30   **Closing session**

Poster

Extracting Biomedical Event Using Feature Selection and Word Representation
Xinyu He, Lishuang Li, Jieqiong Zheng and Meiyue Qin

Overview of the Regulatory Network of Plant Seed Development (SeeDev) Task at the BioNLP Shared Task 2016

Estelle Chaix[*], **Bertrand Dubreucq**[$], **Abdelhak Fatihi**[$], **Dialekti Valsamou**[*,&],
Robert Bossy[*], Mouhamadou Ba[*], Louise Deléger[*] , Pierre Zweigenbaum[&],
Philippe Bessières[*], Loic Lepiniec[$], Claire Nédellec[*]

[*] MaIAGE, INRA, Université Paris-Saclay, 78350 Jouy-en-Josas, France
[$] IJPB, INRA, AgroParisTech, CNRS, Université Paris-Saclay, 78026 Versailles, France
[&] LIMSI, CNRS, Université Paris-Saclay, 91405 Orsay, France
[*]`forename.lastname@jouy.inra.fr`
[$]`forename.lastname@versailles.inra.fr`;[&] `pz@limsi.fr`

Abstract

This paper presents the SeeDev Task of the BioNLP Shared Task 2016. The purpose of the SeeDev Task is the extraction from scientific articles of the descriptions of genetic and molecular mechanisms involved in seed development of the model plant, *Arabidopsis thaliana*. The SeeDev task consists in the extraction of many different event types that involve a wide range of entity types so that they accurately reflect the complexity of the biological mechanisms. The corpus is composed of paragraphs selected from the full-texts of relevant scientific articles. In this paper, we describe the organization of the SeeDev task, the corpus characteristics, and the metrics used for the evaluation of participant systems. We analyze and discuss the final results of the seven participant systems to the test. The best F-score is 0.432, which is similar to the scores achieved in similar tasks on molecular biology.

1 Introduction

Since its first edition in 2009, BioNLP Shared Task (BioNLP-ST) organizes information extraction (IE) tasks from scientific literature with a focus on molecular mechanisms with the aim to promote advances in IE research in the biomedical domain. The SeeDev task is the first task on event extraction about molecular biology of plants. It gives an opportunity for the BioNLP community to evaluate the reusability of methods, to characterize the peculiarities of IE for the plant biology domain and to develop dedicated approaches. For this purpose, we manually annotated a new corpus of scientific papers selected for their relevance to the topic. We propose to the participants to extract text-bound events that involve biological entities provided as input. The performances of the systems are evaluated by standard measures through the comparison of their predictions to the reference annotations.

2 Context

Seeds are the main vectors for breeding and production of annual field crops. The accumulation of seed storage compounds (*e.g.* sugars, lipids, proteins) is of primary importance for food, feed and industrial uses. Seed development requires the coordinated growth of different tissues that involves complex genetics and environmental regulations (Alberts et al., 2002). A comprehensive understanding of the molecular networks that underlie the regulation of seed development remains a major scientific challenge with important potential impact on fundamental research, agriculture and industry.

The SeeDev task of BioNLP Shared Task 2016 focuses on the accumulation of reserves in the seed of the model plant, *Arabidopsis thaliana* (*Ath*), for which research on regulatory networks is the subject of a large and active international community (Santos-Mendoza et al., 2008). Most of this knowledge is spread in thousands of articles. As such, this topic constitutes an excellent primer for the development of event extraction methods. The SeeDev corpus should then be largely reusable for the study of other plants and other development phases.

Information Extraction research applied to biology mainly consists in automatic entity extraction, their normalization and event extraction (Ananiadou et al., 2014). The extraction of regulatory network has become one of the most popular tasks in shared tasks in recent years. The increasing

Proceedings of the 4th BioNLP Shared Task Workshop, pages 1–11,
Berlin, Germany, August 13, 2016. ©2016 Association for Computational Linguistics

complexity of the event scheme over the years is driven by the significant scientific advances in IE and the increasing need for computational models in bioinformatics and systems biology. In 2005, the objective of the *Learning Language in Logic* challenge (LLL'05) was the extraction of gene interactions between proteins and genes with the goal of reconstructing bacterial regulatory networks (Nédellec, 2005). The diversity of the biological events (molecular, physiological) and entities (genes, proteins, families, sites, environmental factors and phenotypes) has continuously increased over the time together with the variety of the biological mechanisms studied. These mechanisms range from detailed networks as in *Bacteria Interaction* (Bossy et al., 2012) and *Gene Regulation Network* (Bossy et al., 2015) tasks, signaling pathways as in *GENIA* task (Kim et al., 2013a) and metabolism to diseases as in *Pathway Curation* (*PC*) and *Cancer Genetics* (*CG*) tasks (Pyysalo et al., 2015). Their extraction from text makes an increasing use of existing standards, nomenclatures and ontologies such as Gene Ontology that facilitates the integration of the text mining results into larger knowledge bases and bioinformatics applications (*e.g.* GRO task (Kim et al., 2013b)) or OntoBiotope (*e.g. Bacteria Biotope* task (Bossy et al., 2015)).

The SeeDev task brings a new application domain, plant development biology, with similar goals and representation as previous IE shared tasks on biological event extraction. This new application domain has required the design of a new knowledge model for the representation of the events, a manually annotated corpus and new metric that accounts for the varying importance of the event arguments.

We refer to the SeeDev task knowledge model as *Gene Regulatory Network for Arabidopsis* (GRNA). GRNA meets the usual constraints of manual annotation of texts (*e.g.* biological relevance and computational tractability), and of automatic annotation by IE methods (*e.g.* learnability from training examples). We have also taken into account the expected use of GRNA for the indexing and retrieval of textual events and experimental data in a unified representation, the modeling of other plant systems, and also the integration of text knowledge with knowledge derived from experimental data.

SeeDev corpus is composed of paragraphs from a selection of recent full-text scientific papers about molecular biology of seed development.

3 Task Description

The SeeDev Task consists in two subtasks (1) *SeeDev-binary* on binary relation extraction and (2) *SeeDev-full* on full event extraction. The *SeeDev-binary* subtask has been conceived as a first step towards the extraction of full n-ary events, which is of interest for plant biology. Both subtasks share the same GRNA model and the same document set with different annotation sets. The two annotations sets contain binary relations and events respectively. The annotation set of *SeeDev-binary* has been computed from the annotation set of *SeeDev-full* through the application of formal transformation rules.

3.1 Knowledge Representation

The GRNA model defines 16 entity types (Figure 1) and 21 event types (Table 1). They are classified into categories and subcategories for readability purpose.

Molecule:
 DNA: *Gene, Gene_Family, Box, Promoter*
 DNA product : *RNA, Protein, Protein_Family, Protein_Complex, Protein_Domain*
  Hormone: *Hormone*
 Dynamic Process: *Regulatory_Network, Pathway*
  Context: *Tissue, Development_Phase, Genotype, Environmental_Factor*

Figure 1. SeeDev entity types.

The *Molecule* category includes molecules that are directly involved in regulation, such as *Hormone* that plays a critical role in plant growth, and *Protein Domain* and DNA regions (*Box, Promoter*) for the representation of physical binding events. Protein and gene families are also important entities because they are mentioned as actors of the regulations in some papers without more precision on the exact molecule. The *Dynamic Process* category is defined by two broad entity types, *Regulatory Network* and *Metabolic pathway*, with the purpose of keeping the complexity of the extraction task tractable. Moreover, the distinction in the SeeDev corpus between specific kinds of networks or pathways would have been difficult, if not impossible because the authors themselves remain vague.

Relation Name	Definition	#	Train	Dev	Test	Total
Regulation		1731	46%	22%	31%	48%
Regulates Accumulation (Regulation Of Accumulation)	A Molecule, Dynamic Process or Context regulates the accumulation of a Functional Molecule (in particular, [*Protein*], [*RNA*], [*Hormone*]).	81	44%	36%	20%	2%
Regulates Development Phase (Regulation Of Development Phase)	A Molecule, Dynamic Process or Context regulates the activity of a Development phase.	242	44%	24%	32%	7%
Regulates Expression (Regulation Of Expression)	**A Molecule, Dynamic Process or Context regulates the expression of a DNA entity. DNA entity includes [*Promoter*] and [*Box*].**	**450**	**45%**	**25%**	**31%**	**13%**
Regulates Molecule Activity (Regulation Of Molecule Activity)	An Agent (Molecule, Dynamic Process or Context) regulates the activity of a Molecule, such as [*Protein*].	25	64%	0%	36%	1%
Regulates Process (Regulation Of Process)	**A Molecule, Dynamic Process or Context regulates the activity of a Dynamic Process.**	**904**	**48%**	**20%**	**32%**	**25%**
Regulates Tissue Development (Regulation Of Tissue Development)	A Molecule, Dynamic Process or Context regulates the activity of a Tissue Development.	29	31%	31%	38%	1%
Function		257	42%	28%	30%	7%
Is Involved In Process (Involvement In Process)	A Molecule is involved *in* a Dynamic Process.	55	42%	36%	22%	2%
Transcribes Or Translates To (Transcription Or Translation)	A DNA entity encodes for a RNA (Transcription) or a RNA entity encodes a Protein (Translation). Often, reference is made to the gene encoding the protein, without mention of the RNA.	54	46%	24%	30%	2%
Is Functionally Equivalent To* (Functional Equivalence)	A Molecule, Dynamic Process or Context is compared to a similar entity.	148	41%	26%	33%	4%
Interaction		264	46%	21%	33%	7%
Interacts With (Interaction)	A molecule interacts with another molecule.	148	42%	22%	36%	4%
Binds To (Binding)	A functional molecule physically binds to a molecule.	116	52%	21%	28%	3%
Where and When		704	45%	23%	32%	20%
Exists At Stage (Presence At Stage)	A Molecule is present *during* a Developmental phase.	33	45%	24%	30%	1%
Exists In Genotype (Presence In Genotype)	**A Molecule or Element is present *in* a Genotype**	**377**	**45%**	**21%**	**34%**	**11%**
Occurs During (Occurrence During)	A Process occurs *during* a Developmental Phase.	30	27%	33%	40%	1%
Occurs In Genotype (Occurrence In Genotype)	A Process occurs *in* a Genotype	48	38%	33%	29%	1%
Is Localized In (Localization)	A Molecule is found in a Tissue	216	50%	22%	29%	6%
Composition and Membership		532	44%	22%	34%	15%
Composes Primary Structure (Primary Structure Composition)	A specific sequence of nucleotide is found in a DNA entity.	51	39%	29%	31%	1%
Composes Protein Complex (Protein Complex Description)	A specific DNA product is found in a Protein complex.	19	84%	0%	16%	1%
Has Sequence Identical To* (Sequence Identity)	A Molecule, Dynamic Process or Context is compared to a similar Molecule, Dynamic Process or Context.	126	49%	16%	35%	4%
Is Member Of Family (Family Membership)	A DNA, RNA or Protein belongs to another DNA, Product or Factor. Used between entities of the same nature to denote members of a set.	230	39%	24%	37%	6%
Is Protein Domain Of (Protein Domain Composition)	A specific Protein Domain is found in an amino acid sequence.	106	43%	27%	29%	3%
Specific to Binary scheme		87	51%	26%	23%	2%
Is Linked To*	Used to derive binary relations from n-ary events: it relates optional and main arguments of n-ary events.	87	51%	26%	23%	2%
Total		3575	46%	23%	32%	100%

Table 1: Definition of relations and example distribution in SeeDev *Binary* subtask. Event names are into brackets. (Event arguments are ordered, except events marked with *.)

N-ary representation : Binding							
	Mandatory arguments		Optional arguments				
Role	Functional Molecule	Molecule	Tissue	Developmental Stage	Organism Genotype	Environmental Factor	Hormone
Signature	*RNA, Protein, Protein Family, Protein Complex, Protein Domain, Hormone*	*Gene, Gene Family, Box, Promoter, RNA, Protein, Protein Family, Protein Complex, Protein Domain,*	*Tissue*	*Development Phase*	*Genotype*	*Environmental Factor*	*Hormone*
Binary representation : Binds_to							

Figure 2: Representation of *Binds_to* and *Binding* relation, with mandatory and optional arguments.

Arg 1 \ Arg 2	Gene	Gene Family	Box	Promoter	RNA	Protein	Protein Family	Protein Complex	Protein Domain	Hormone	Regulatory Network	Metabolic pathway	Genotype	Tissue	Development Phase	Environmental Factor
Gene	5	6	3	3	3	5	5	5	3	3	3	3	1	1	1	1
Gene Family	5	6	3	3	3	5	5	5	3	3	3	3	1	1	1	1
Box	3	3	6	4	2	4	4	4	2	3	3	3	1	1	1	1
Promoter	3	3	4	6	2	4	4	4	2	3	3	3	1	1	1	1
RNA	3	3	3	3	6	6	6	6	4	4	3	3	1	2	2	1
Protein	4	4	4	4	4	7	8	6	3	4	3	3	1	2	2	1
Protein Family	4	4	4	4	4	7	8	6	3	4	3	3	1	2	2	1
Protein Complex	4	4	4	4	4	5	5	8	3	4	3	3	1	2	2	1
Protein Domain	4	4	4	4	4	6	6	7	6	4	3	3	1	2	2	1
Hormone	3	3	3	3	3	4	4	4	2	6	3	3	1	2	2	1
Regulatory Network	2	2	2	2	2	3	3	3	1	3	4	2	1	2	2	1
Metabolic pathway	2	2	2	2	2	3	3	3	1	3	2	4	1	2	2	1
Genotype	1	1	1	1	1	2	2	2	0	2	1	1	3	1	1	0
Tissue	1	1	1	1	1	2	2	2	0	2	1	1	1	3	1	0
Development Phase	1	1	1	1	1	2	2	2	0	2	1	1	1	1	3	0
Environmental Factor	2	2	2	2	2	3	3	3	1	3	2	2	1	2	2	3

Figure 3: Number of relation type by pairs of argument types.

The conditions in which the regulations occur represent critical information about the event context. The entity types represent spatial conditions (*Tissue*), temporal conditions (*Development phase*), the organism, which is genetically modified or not (*Genotype*), and the environmental factors (biotic and abiotic external conditions). The entities in the corpus are denoted by individual words or by sets of words that may be discontinuous.

The 21 GRNA event types are grouped in 6 sets, according to their biological role (Table 1). The *Regulation*, *Function* and *Interaction* categories are central for the description of the biological mechanisms. *Where and When* event types represent the context of the mechanisms, whilst *Composition and Membership* events allow to finely represent relations among the biological entities. Some of the event types, *e.g. Regulates Expression / Process / Molecule Activity* are very similar to those of other molecular biology IE event schemes such as the ones of *GENIA* (Kim et al., 2013a), *Cancer Genetics* (Pyysalo et al., 2015) and *Arabidopsis Leaf Growth* (*LG*) (Szakonyi et al., 2015). Other GRNA event types are specific to biological development, *e.g. Regulates Development Phase / Tissue Development* or to the storage process, *e.g. Regulates Accumulation*. The

LG model of Szakonyi et al. (2015) dedicated to *Ath* does not include plant or development specific events to be reused in GRNA. Protein modification and metabolism in GENIA and PC tasks and regulation of phenotype in *LG*, were not relevant for the SeeDev corpus but will be addressed in priority in further extensions of GRNA.

The first column of Table 1 displays the binary relation names of *SeeDev-binary* subtask and the n-ary event names of *SeeDev-full* subtask in brackets, with their definition in column two. N-ary events have two mandatory arguments and up to five optional arguments: *Tissue, Developmental Stage, Organism, Genotype, Environmental Factor*, and *Hormone*.

Furthermore, n-ary events may have a negation modality. Participants are provided with text documents, gold entity annotations, and the detailed signatures of each event, *i.e.* the list of allowed types per slot. Figure 2 gives, for example, the *Binding* event signature.

The use of a strongly typed model facilitates the event prediction because it drastically reduces the number of event candidates given the types of the arguments. Figure 3 shows the number of relation types per pair of argument types. For example the argument pair (*Arg1: Development_Phase / Arg2: Protein_Domain*) does not accept any relation type; whereas the pair (*Arg1: Protein / Arg2: Protein_Family*) may be involved into 8 different relations. The formal specification of event signatures drastically reduces the exploration space of possible events.

3.2 Sub-Task 1: SeeDev Binary Relation Extraction

The goal of *SeeDev-binary* is the extraction of binary relations of 22 different types without modality (no negation) as described in Table 1. The *Is_Linked_To* relation is computed from the n-ary events, it links mandatory arguments to optional arguments. Figure 4.a gives an example of *SeeDev-binary* annotation with 3 different relations.

3.3 Sub-Task 2: SeeDev Full Relation Extraction

SeeDev-full aims at extracting n-ary events where the number of arguments ranges from two to eight, plus a negation modality. There are three arguments in average. There is no trigger word in SeeDev event representation. Events relate

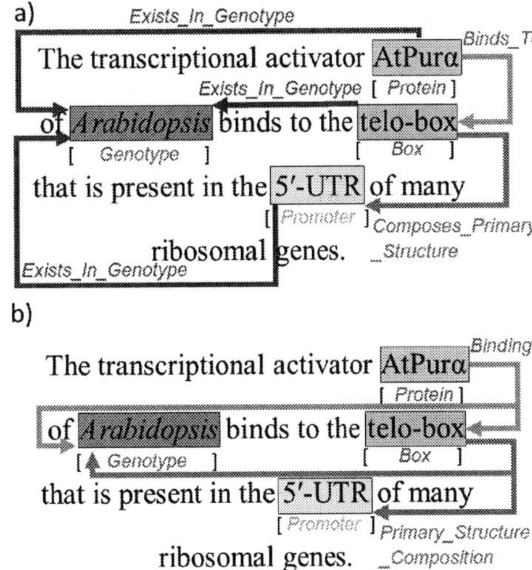

Figure 4: Examples of an annotated sentence in (a) *SeeDev-binary* task and (b) *SeeDev-full* task

either entities or other events. Figure 4.b gives an example of a *Binding* event with a *Genotype* argument. In the binary version (Figure 4.a), the *Genotype* becomes a mandatory argument of one of the *Exists_In_Genotype* relations.

4 Corpus Description

The SeeDev corpus is a set of 86 paragraphs from 20 full-text articles, selected by plant biology experts, about seed development in *Arabidopsis thaliana*. Table 2 summarizes the SeeDev corpus statistics and data distribution in the Training, Development and Test sets.

	#	Train	Dev	Test
Documents	20	90%	75%	80%
Paragraphs	87	45%	22%	33%
Words	44,857	45%	23%	33%
Entities	7,082	46%	23%	31%
Events	2,583	45%	23%	32%
Relations	3,575	46%	23%	32%

Table 2. SeeDev corpus statistics.

Paragraphs of the same document may be distributed into different sets. The "Documents" row indicates the proportion of documents represented in the set. The SeeDev corpus is smaller than other BioNLP-ST corpora, *e.g.* a fifth of *Cancer Genetics* corpus and a third of *GENIA* corpus. The manual annotation of the SeeDev corpus required a high level of expertise that do not allow for a large corpus, as in many specific domains of Life

Science. We identify small dataset processing as a challenge to overcome by information extraction tools.

Table 1 details the distribution of instances per relation type in the training, development and test sets of the *SeeDev-binary* task. The distribution was balanced between the three data sets so that the test set would represent approximately a third of the annotations for each group of relations. The most frequent relations are *Regulation* with 48% of annotations, which corresponds to what is expected given the corpus domain. The three relations *Regulate Expression*, *Regulates Process* and *Exist in Genotype,* highlighted in Table 1, account for half of the total, whilst seven of the relations are relatively infrequent with 1% of the total.

5 Annotation Methodology

We have successively refined the annotation scheme of GRNA during the annotation process. We have defined an initial annotation scheme according to our expertise in *A. thaliana* seed development and in BioNLP task definition, starting from the GRN model (Bossy et al., 2015).

The scheme was improved through several iterations of manual annotations and collective discussions until it met the requirements, *i.e.* it allowed unambiguous, consistent, readable and detailed formal annotations. Together with the scheme, a very precise guideline document (Chaix et al., 2016) was produced that details the annotation principles for each entity and event type, and provides many examples and counter-examples.

The relevant paragraphs of the corpus were chosen by the biologists, mostly from the abstract, introduction, result and discussion sections. A team of three experts in seed development and two bioinformaticians has manually annotated the corpus following the guidelines by using the AlvisAE Annotation Editor (Papazian et al., 2012) in accordance with the final version of the scheme.

5.1 Automatic Annotation

Rigid designators of named entities, such as *Gene*, *Protein*, *Tissues*, and *Developmental Phases* were automatically pre-annotated with the AlvisNLP pipeline using relevant *Ath* databases (*e.g.* TAIR[1]) and customized lexicons. The goal of automatic

pre-annotation was to speed-up the manual annotation process. The evaluation of the automatic annotation compared to the gold standard annotation shows a F-score equal to 0.41, with a high precision (0.89) and low recall (0.26) due to a lack of relevant lexicon for most entity types.

5.2 Manual Annotation

The manual annotation has been achieved in four successive phases in order to both save expert time and achieve a high quality annotation. First, a bioinformatician who is not a specialist of *Ath* annotated all the entities of the corpus. The evaluation of the manual annotation of the entities compared to the gold standard annotation yielded a high 0.93 F-score with balanced Recall and Precision, 0.93 and 0.95 respectively.

Then *Ath* experts revised the entity annotations and annotated the events of the corpus in a double-blind manner. Thanks to the manual pre-annotation of entities, they could focus on events which require more expertise. Next, the annotators together with the bioinformatician used the AlvisAE conflict resolution functionality to build a consensus. Finally, the bioinformatician carefully checked the compliance of each annotation to the guidelines to produce the gold annotation set.

To evaluate the inter-annotator agreement, we measured the F-score between the annotation set of each annotator (referred to as A and B) and the consensus annotation set (*i.e.* gold annotations) (Table 3). The differences between the individual annotators vary according to the event types. The recall measure of the annotations of events with arguments of Process type without regulation (*Is Involved In Process*) and events with Genotype arguments (*Exists In Genotype, Occurs In Genotype*) is lower.

Mistyping *Regulates Accumulation* was frequent because this event is easily confused with *Regulates Molecule Activity*. Annotations from annotator B are closer to the reference annotation, but the examination of the union of both annotation sets shows that annotator B missed events that were well annotated by A. The 0.724 F-score of the union of A and B annotation sets is quite high. The last step of the SeeDev corpus construction is the adjudication between the two annotators with a third person as external referee. It was an essential step to avoid event oversight.

---

[1]The Arabidopsis Information Resource `http://arabidopsis.org/`

Annotator	F1	Recall	Precision
A	0.548	0.417	0.798
A (T)	+0.048	+0.031	+0.058
B	0.653	0.575	0.754
B (T)	+0.069	+0.071	+0.080
A U B	0.724	0.720	0.728
A U B (T)	**+0.045**	**+0.045**	**+0.045**

Table 3: Evaluation of the inter-annotator agreement by comparing each annotator output to the reference annotation. (T) indicates the gain if relation types are ignored. A U B denotes the union of annotations from annotators A and B.

6 Evaluation Procedure

6.1 Shared Task Organization

As for previous challenges, BioNLP-ST 2016 provides resources and information to the participants through the BioNLP-ST website[2] and mailing lists. The schedule of the SeeDev task follows the usual principles of BioNLP-ST tasks, it can be found on dedicated pages.

We provided state-of-art automatic NLP analysis as supporting resources with the purpose to speed-up the participant system development. Nine tools were selected and applied to the training, development and test sets: POS tagger (*GENIA Tagger* (Tsuruoka et al., 2005)), parsers (*Stanford Parser* (Manning, 2003) *Enju* (Miyao and Tsujii, 2008) *C&C CCG Parser* (Clark and Curran, 2007)), term extractor (*BioYaTeA* (Golik et al., 2013)) named entity recognizers (*Stanford NER* (Finkel et al., 2005) *LINNAEUS* (Gerner et al., 2010) *SR4GN* (Wei et al., 2012)) and tokenizer and sentence splitter (*AlvisNLP suite* (Ba and Bossy, 2016)).

Community web tools (forum, FAQ and mailing list) have been made available on the website with the purpose to federate the community that participates to the challenge. In this way participants could interact with the task organizers and with other participants.

Furthermore, participants could evaluate their predictions through an online evaluation service. During the training phase it was restricted to the evaluation on training and development sets. The service allows now to evaluate predictions on the test set and will remain open. For the first time in BioNLP-ST, participants could also keep track

of the performance of various experiments through the same online service. Thus, participants could follow and compare their results and competing team results. The recorded submissions were kept anonymous to other participants. The aim of this tool was to ease the interpretation of the scores and to assist participants in the development-test cycles.

6.2 Evaluation Metrics

The evaluation measures of the participant system results are computed through the comparison of predicted events against reference corpus events. In *SeeDev-binary* the participants had to predict relations between entities given as input. This task can be viewed as a classification task of all pairs of entities. Thus, we evaluate submissions with Recall, Precision and F-score. Submissions were ranked by F-score, however we also provided alternate evaluations in order to assess the strengths of each submission for each relation type separately, for each broad category of relations separately and without taking into account the relation types.

We also designed a measure for *SeeDev-full* task evaluation that is permissive for optional arguments. The evaluation is detailed on the task web site and is available through the online evaluation service to the benefit of teams that will bravely tackle this task.

7 Results

7.1 Participating Systems

Seven teams from 4 continents submitted their results to the test of the SeeDev binary task that are: *DUTIR* (Dalian University of Technology, China), *LIMSI* (CNRS, France), *LitWay* (Xidian University, China), *ULisboa* (LaSIGE, Universidade de Lisboa, Portugal), *UniMelb* (University of Melbourne, Australia), *VERSE* (University of British Columbia, Canada) and *UTS* (University of Turku, Finland).

Their main background domains are Bioinformatics, Machine Learning, Natural Language Processing and Biology according to their responses to a survey.

Table 4 summarizes the scores obtained by the participant systems ranked by F1-score (detailed results are available on the SeeDev site). The results of the *DUTIR* system are not displayed because they experienced a last minute hitch

[2]BioNLP-ST website `http://2016.bionlp-st.org/tasks/seedev`

and ranked last. *LitWay* from Xidian University achieves the best F1-score (0.432), 0.068 points higher than the second team and 0.177 points higher than the lowest score at 0.255. The two systems that ranked first achieved a balanced recall and precision, while the four others favored recall over precision (*VERSE, LIMSI*), or the reverse (*UTS, ULISBOA*). *VERSE* obtained the best recall and *UTS* the best precision.

Participant	F1	Recall	Precision
LitWay	0.432	0.448	0.417
UniMelb	0.364	0.386	0.345
VERSE	0.342	0.458	0.273
UTS	0.335	0.245	0.533
ULISBOA	0.306	0.256	0.379
LIMSI	0.255	0.318	0.212

Table 4: Evaluation scores of the SeeDev binary task ranked by F- score.

The best F1-scores are very similar to the ones achieved by participants of previous shared tasks on regulation event extraction around 50% (*e.g.* GRN, CG, PC), which is over what could be expected given the complexity and the novelty of the task and the variability of the example distribution among the events.

As shown by Table 5, the detailed scores per relation exhibit a high variability. Some relations were difficult to predict (*e.g. Regulates Tissue Development, Regulates Molecule Activity, Occurs During*) while others were well-predicted (*e.g. Composes Primary Structure* with a maximum F1-score of 0.67).

As usual in such corpus, the analysis of the results shows that the causes are multifactorial, we hypothesize that the number of training examples combined with the regularity of the descriptions and the constraints imposed by the event signature are critical. For instance, the *Composes Primary Structure* relation has only 51 examples, but it links entities from a restricted range of types, which makes it easier to predict (0.67 best F1-score). However, other relations such as *Regulates Expression with* a high number of examples (450 examples), inter sentence occurrences (23) and a wide range of argument types (4 types for the first argument and 16 for the second) were poorly predicted (0.39 best F1-score).

The scores of most of the systems remain unchanged when the dataset is restricted to the

Relation	Best F1 score	System
All Relations	**0.432**	**LitWay**
Where and When	**0.142**	**LitWay**
Exists_At_Stage	0.167	ULISBOA
Exists_In_Genotype	0.492	LitWay
Occurs_During	0	-
Occurs_In_Genotype	0.167	VERSE
Is_Localized_In	0.450	LitWay
Function	**0.255**	**ULISBOA**
Is_Involved_In_Process	0	-
Transcribes_Or_Translates_To	0.343	VERSE
Is_Functionally_Equivalent_To	0.708	LitWay
Regulation	**0.416**	**LitWay**
Regulates_Accumulation	0.316	UniMelb
Regulates_Development_Phase	0.376	UniMelb
Regulates_Expression	0.386	UniMelb
Regulates_Molecule_Activity	0	
Regulates_Process	0.504	LitWay
Regulates_Tissue_Development	0	-
Composition_MemberShip	**0.490**	**LitWay**
Composes_Primary_Structure	0.667	LIMSI
Composes_Protein_Complex	0.500	UTS
Has_Sequence_Identical_To	0.867	LitWay
Is_Member_Of_Family	0.534	LitWay
Is_Protein_Domain_Of	0.438	LitWay
Interaction	**0.303**	**UniMelb**
Interacts_With	0.286	UniMelb
Binds_To	0.310	VERSE
Is_Linked_To	0.154	VERSE

Table 5: Best F1-score per relation and per category of relation.

relations that occur in a single sentence. The difference of the results obtained for intra-sentence dataset are less than 1 point, except for *Limsi* that gains 0.056 points; indeed, *Limsi* is the only team that attempts to predict inter-sentence relations whereas all other participant systems predicted only intra-sentence relations. Given the proportion of inter-sentence relations in the test set (4%), the penalty of ignoring them could have been considered as bearable.

In order to assess the difficulty to predict the correct relation type, we computed the F-scores when considering the category of the relations instead of the actual type (first line per category in bold and italic in Table 5). This did not yield a significant improvement although some participants were able to successfully predict events in categories with high biological relevance, such as the *Regulation* category (*Litway* F1: 0.416) and the *Interaction* category (*UniMel* F1: 0.303).

7.2 Systems Description and Result Discussion

All teams used supervised machine-learning approaches (Table 6). Five systems used support vector machines (SVM) and two systems were based on different algorithms, namely *maximum entropy* (MaxEnt) (*LIMSI*) and *convolutional neural network* (*DUTIR*).

Participant	General method
LitWay	Hand crafted patterns + SVM
UniMelb	SVM + Bayes classifiers
VERSE	Linear SVM
ULISBOA	SVM kernel based
UTS	SVM multi-classification
LIMSI	Bag of words
DUTIR	Convolutional neural network

Table 6: General methods of the participants

SVM are widely used for information extraction tasks, because they are powerful versatile classifiers. SVM are kernel-based and there are several existing kernels available (Zelenko et al., 2003) adapted to different object representations. For instance, dependency-path kernels (Bunescu and Mooney, 2005; Airola et al., 2008) handle candidates represented as syntactic dependency paths. Moreover, the usual feature selection methods can be handled by kernels that work on vectorial representations. MaxEnt and neural networks are also popular algorithms in information extraction tasks (McCallum et al., 2000). The most notable characteristic of the best performing system, *LitWay*, is that it combines supervised machine learning for the prediction of a selection of event types with hand-crafted rules for the prediction of other types.

All teams used token segmentation, sentence splitting and token normalization (stemming, lemmatization, POS-tagging). Four teams, among which the three top ranking also used deep syntactic parsing, which confirms that parsing is a powerful pre-processing step for information extraction. Finally, the *LitWay* system also designed features based on word embedding which is a novelty in the BioNLP-ST.

8 Conclusion

We have described the SeeDev task that we have designed with the goal to promote progress in information extraction in the field of plant development and more precisely plant regulatory networks. Two sub-tasks were proposed with increasing levels of complexity, *SeeDev-binary* on binary relations and *SeeDev-full* on events.

The lack of participation to *SeeDev-full* shows that the extraction of n-ary events with optional arguments remains challenging.

Seven teams from different countries participated in the *SeeDev-binary* task with different approaches. The results are very promising, given the novelty of the task and the complexity of the model. The best F-score, 0.432, is close to what has been previously obtained in similar IE tasks on molecular biology.

The good results achieved by hybrid methods using machine learning and handcraft patterns show that efficient adaptation of generic methods to the task could rely not only on machine learning, but also on alternative approaches. This observation may also be true for the extraction of n-ary events from binary relations where rewriting rules may complement machine learning methods. This may be particularly appropriate for relatively small corpora as SeeDev, which belongs to a domain where a trade-off has to be found between the time needed for the training corpus annotation and the time needed for the manual development of dedicated rules for the IE method.

Acknowledgments

This work was supported by the French National Institute for Agricultural Research (Inra), the Center for Data Science (CDS), the OpenMinTeD project (EC/H2020-EINFRA 654021) and the Institute for Living Systems (IMSV), funded by the IDEX Paris-Saclay ANR-11-IDEX-0003-02. The IJPB benefits from the support of the Labex Saclay Plant Sciences- SPS (ANR-10-LABX-0040-SPS).

References

Antti Airola, Sampo Pyysalo, Jari Björne, Tapio Pahikkala, Filip Ginter, and Tapio Salakoski. 2008. All-paths graph kernel for protein-protein interaction extraction with evaluation of cross-corpus learning. *BMC bioinformatics*, 9(11):1.

B Alberts, A Johnson, J Lewis, P Walter, M Raff, and K Roberts. 2002. Molecular biology of the cell 4th edition: International student edition.

Sophia Ananiadou, Paul Thompson, Raheel Nawaz, John McNaught, and Douglas B Kell. 2014. Event-based text mining for biology and functional genomics. *Briefings in functional genomics*, page elu015.

Mouhamadou Ba and Robert Bossy. 2016. Interoperability of corpus processing work-flow engines: the case of alvisnlp/ml in openminted. In Richard Eckart de Castilho, Sophia Ananiadou, Thomas Margoni, Wim Peters, and Stelios Piperidis, editors, *Proceedings of the Workshop on Cross-Platform Text Mining and Natural Language Processing Interoperability (INTEROP 2016) at LREC 2016*, pages 15–18, Portoroz, Slovenia, May. European Language Resources Association (ELRA).

Robert Bossy, Julien Jourde, Alain-Pierre Manine, Philippe Veber, Erick Alphonse, Maarten Van De Guchte, Philippe Bessières, and Claire Nédellec. 2012. Bionlp shared task-the bacteria track. *BMC bioinformatics*, 13(11):1.

Robert Bossy, Wiktoria Golik, Zorana Ratkovic, Dialekti Valsamou, Philippe Bessières, and Claire Nédellec. 2015. Overview of the gene regulation network and the bacteria biotope tasks in bionlp'13 shared task. *BMC bioinformatics*, 16(10):1.

Razvan C Bunescu and Raymond J Mooney. 2005. A shortest path dependency kernel for relation extraction. In *Proceedings of the conference on human language technology and empirical methods in natural language processing*, pages 724–731. Association for Computational Linguistics.

Estelle Chaix, Bertrand Dubreucq, Dialekti Valsamou, Abdelhak Fatihi, Louise Deléger, Robert Bossy, Pierre Zweigenbaum, Philippe Bessières, Loic Lepiniec, and Claire Nédellec. 2016. Annotation guidelines bionlp-st 2016 seedev task. Technical report, INRA.

Stephen Clark and James R Curran. 2007. Wide-coverage efficient statistical parsing with ccg and log-linear models. *Computational Linguistics*, 33(4):493–552.

Jenny Rose Finkel, Trond Grenager, and Christopher Manning. 2005. Incorporating non-local information into information extraction systems by gibbs sampling. In *Proceedings of the 43rd Annual Meeting on Association for Computational Linguistics*, pages 363–370. Association for Computational Linguistics.

Martin Gerner, Goran Nenadic, and Casey M Bergman. 2010. Linnaeus: a species name identification system for biomedical literature. *BMC bioinformatics*, 11(1):1.

Wiktoria Golik, Robert Bossy, Zorana Ratkovic, and Claire Nédellec. 2013. Improving term extraction with linguistic analysis in the biomedical domain. In *Proceedings of the 14th International Conference on Intelligent Text Processing and Computational Linguistics (CICLing13), Special Issue of the journal Research in Computing Science*, pages 24–30.

Jin-Dong Kim, Yue Wang, and Yamamoto Yasunori. 2013a. The genia event extraction shared task, 2013 edition-overview. In *Proceedings of the BioNLP Shared Task 2013 Workshop*, pages 8–15. Association for Computational Linguistics.

Jung-Jae Kim, Xu Han, Vivian Lee, and Dietrich Rebholz-Schuhmann. 2013b. Gro task: Populating the gene regulation ontology with events and relations. In *Proceedings of the BioNLP Shared Task 2013 Workshop*, pages 50–57.

DKCD Manning. 2003. Natural language parsing. *Advances in neural information processing systems*, 15:3.

Andrew McCallum, Dayne Freitag, and Fernando CN Pereira. 2000. Maximum entropy markov models for information extraction and segmentation. In *Icml*, volume 17, pages 591–598.

Yusuke Miyao and Jun'ichi Tsujii. 2008. Feature forest models for probabilistic hpsg parsing. *Computational Linguistics*, 34(1):35–80.

Claire Nédellec. 2005. Learning language in logic-genic interaction extraction challenge. In *Proceedings of the 4th Learning Language in Logic Workshop (LLL05)*, volume 7, pages 1–7. Citeseer.

Frédéric Papazian, Robert Bossy, and Claire Nédellec. 2012. Alvisae: a collaborative web text annotation editor for knowledge acquisition. In *Proceedings of the Sixth Linguistic Annotation Workshop*, pages 149–152. Association for Computational Linguistics.

Sampo Pyysalo, Tomoko Ohta, Rafal Rak, Andrew Rowley, Hong-Woo Chun, Sung-Jae Jung, Sung-Pil Choi, Jun'ichi Tsujii, and Sophia Ananiadou. 2015. Overview of the cancer genetics and pathway curation tasks of bionlp shared task 2013. *BMC bioinformatics*, 16(10):1.

Monica Santos-Mendoza, Bertrand Dubreucq, Sébastien Baud, François Parcy, Michel Caboche, and Loïc Lepiniec. 2008. Deciphering gene regulatory networks that control seed development and maturation in arabidopsis. *The Plant Journal*, 54(4):608–620.

Dóra Szakonyi, Sofie Van Landeghem, Katja Baerenfaller, Lieven Baeyens, Jonas Blomme, Rubén Casanova-Sáez, Stefanie De Bodt, David Esteve-Bruna, Fabio Fiorani, Nathalie Gonzalez, et al. 2015. The knownleaf literature curation system captures knowledge about arabidopsis leaf growth and development and facilitates integrated data mining. *Current Plant Biology*, 2:1–11.

Yoshimasa Tsuruoka, Yuka Tateishi, Jin-Dong Kim, Tomoko Ohta, John McNaught, Sophia Ananiadou, and Junichi Tsujii. 2005. Developing a robust part-of-speech tagger for biomedical text. In *Panhellenic Conference on Informatics*, pages 382–392. Springer.

Chih-Hsuan Wei, Hung-Yu Kao, and Zhiyong Lu. 2012. Sr4gn: a species recognition software tool for gene normalization. *PloS one*, 7(6):e38460.

Dmitry Zelenko, Chinatsu Aone, and Anthony Richardella. 2003. Kernel methods for relation extraction. *Journal of machine learning research*, 3(Feb):1083–1106.

Overview of the Bacteria Biotope Task at BioNLP Shared Task 2016

[1]**Louise Deléger**, [1]**Robert Bossy**, [1]**Estelle Chaix**, [1]**Mouhamadou Ba**, [1,2]**Arnaud Ferré**,
[1]**Philippe Bessières**, [1]**Claire Nédellec**
[1]MaIAGE, INRA, Universit Paris-Saclay, 78350 Jouy-en-Josas, France
[2]LIMSI, CNRS, Universit Paris-Saclay, 91405 Orsay, France
`firstname.lastname@jouy.inra.fr`

Abstract

This paper presents the Bacteria Biotope task of the BioNLP Shared Task 2016, which follows the previous 2013 and 2011 editions. The task focuses on the extraction of the locations (biotopes and geographical places) of bacteria from PubMed abstracts and the characterization of bacteria and their associated habitats with respect to reference knowledge sources (NCBI taxonomy, OntoBiotope ontology). The task is motivated by the importance of the knowledge on bacteria habitats for fundamental research and applications in microbiology. The paper describes the different proposed subtasks, the corpus characteristics, the challenge organization, and the evaluation metrics. We also provide an analysis of the results obtained by participants.

1 Introduction

Since 2009, BioNLP Shared Task is a community-wide effort on the development of fine-grained information extraction methods in biomedicine (Kim et al., 2009; Kim et al., 2011; Nédellec et al., 2013). The tasks provide a sound framework for the comparison and evaluation of the technologies on a manually curated benchmark with the aim to contribute to progress by drawing general lessons from the individual contributions and assessment of the participants. In this paper, we present the third edition of the Bacteria Biotope task that has been first introduced in 2011 with the ambition to use information extraction from scientific documents at a large scale in order to automatically fill knowledge bases (Bossy et al., 2012).

Information about bacteria biotopes (*e.g.*, habitats of bacteria) is critical for studying the interaction and association mechanisms between organisms and their environments from genetic, phylogenetic and ecological points of view. This information is not only highly useful in all fields of applied microbiology such as food processing and safety, health sciences and waste processing, but also in fundamental research (*e.g.*, metagenomics, phylogeography, phyloecology).

Currently, there is no centralized resource gathering the state of knowledge on habitats of bacteria in a comprehensive and normalized way. A large part of this knowledge is scattered in numerous scientific papers and databases, such as genomics databases (*e.g.*, GenBank[1], GOLD[2]), international microorganism culture collections (*e.g.*, ATCC[3], DSMZ[4]), and biodiversity surveys (*e.g.* , GBIF[5]). The information on bacteria biotopes is mostly expressed in free text (*e.g.*, articles or free-text fields of databases) describing very diverse locations (any physical location may be a bacteria habitat) in many different ways. The need for information processing is not only the extraction of habitats and microorganisms relationships from text, but also their normalization with respect to a common referential so that they can be integrated and compared. This need has been acknowledged by previous work on habitat classifications for metagenomic samples (Ivanova et al., 2010), microorganisms (Floyd et al., 2005) and other living organisms (Buttigieg et al., 2013) and text-mining tools for mapping textual descriptions to habitat classification (Pignatelli et al., 2009).

The aim of Bacteria Biotope (BB) task is to provide a framework for the evaluation and compari-

---

[1]`http://www.ncbi.nlm.nih.gov/genbank`
[2]`https://gold.jgi.doe.gov/` (Genomes Online Database)
[3]`http://www.atcc.org/` (American Type Culture Collection)
[4]`https://www.dsmz.de/` (Deutsche Sammlung von Mikroorganismen und Zellkulturen)
[5]`http://www.gbif.org/` (Global Biodiversity Information Facility)

Proceedings of the 4th BioNLP Shared Task Workshop, pages 12–22,
Berlin, Germany, August 13, 2016. ©2016 Association for Computational Linguistics

son of such methods for Bacteria organism habitats. More specifically, the BB task consists in the extraction of bacteria and their locations (habitats or geographical places) from the text, their categorization according to dedicated knowledge sources, and the linking of bacteria to their locations through so-called localization events named "Lives_in". The widely used NCBI taxonomy[6] (Federhen, 2012) is the resource used for Bacteria entity categorization. The OntoBiotope ontology[7], which is dedicated to the description of microorganism habitats, is used for biotope categorization. Previous work has shown the relevance of OntoBiotope for bacteria habitat detection (Ratkovic et al., 2012). The first two editions of the task (Bossy et al., 2012; Bossy et al., 2015) used general-purpose documents, mostly web pages of genomics projects that can be understood by non-specialists. However, scientific literature is the major source of detailed and accurate information on bacteria for biologists. This edition focuses then on scientific paper abstracts from the PubMed database, which offers a twofold advantage, open access and easier readability than full-text. We also introduce this year a new subtask of knowledge base extraction, in which systems are evaluated by measuring how much information content can be extracted from the corpus.

2 Task Description

The BB task involves three types of entities, Bacteria, Habitats and Geographical places. It also involves a single type of event, the *Lives_in* event, which is a relation between two mandatory arguments, the bacterium and the location where it lives, either a *Habitat* or a *Geographical* entity. Figure 1 displays an example of entities and events in the BB task.

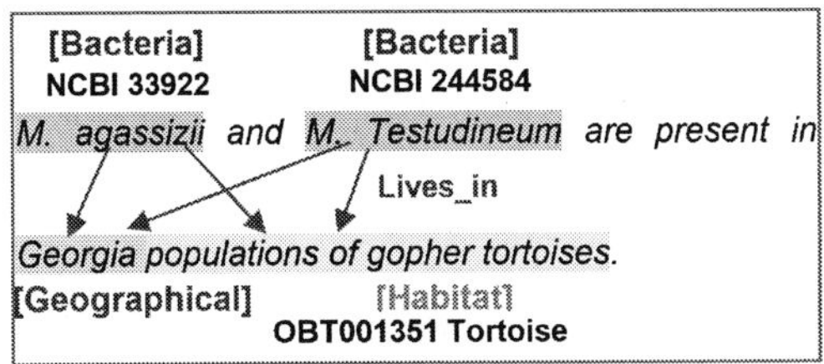

Figure 1: Example of entities and *Lives_in* events in the BB task

[6]http://www.ncbi.nlm.nih.gov/taxonomy
[7]http://2016.bionlp-st.org/tasks/bb2/OntoBiotope_BioNLP-ST-2016.obo

We proposed three subtasks with two modalities each. Each subtask had a plain modality where named entities were given as input, thus participants were not required to perform entity recognition. In the second modality, entities were *not* provided, thus methods had to perform named entity recognition and submissions are partly evaluated on the accuracy of entity boundaries. Our purpose is to assess independently the quality of the methods when dealing with different sub-goals and to assess the impact of predictions made at a given step on the predictions made at the next steps.

2.1 Bacteria and Habitat Categorization

The first subtask focused on the categorization of *Bacteria* and *Habitat* entity occurrences in the text with categories from the NCBI Taxonomy for *Bacteria* and from the OntoBiotope ontology for *Habitat* entities. In the first modality of the subtask (referred to as BB-cat), entity mentions were given and participants had only to perform categorization. In the second modality (BB-cat+ner), systems had to perform bacteria and habitat entity detection as well as categorization.

2.2 Entity and Event Extraction

The second subtask consists in the extraction of *Lives_in* events among *Bacteria*, *Habitat* and *Geographical* entities. In the BB-event modality, entity mentions were given and participant systems only had to perform event extraction. In the BB-event+ner modality, systems had to perform *Bacteria*, *Habitat* and *Geographical* entity recognition as well as event extraction.

2.3 Knowledge Base Extraction

The third subtask aims at building a knowledge base using information extracted from the corpus. The knowledge base is composed of the set of distinct *Lives_in* events between categorized *Bacteria* and *Habitats*. This subtask can be seen as a combination of the entity categorization and event extraction subtasks. In contrast with the two previously described subtasks, this task does not evaluate text-bound annotations. All pieces of information extracted from the text are gathered and merged into a single knowledge base, without duplicate events. The focus of this task is the knowledge itself (which types of bacteria and habitat are linked through a *Lives_In* event) and not the individual text-bound annotations (where *Lives_In* events are marked precisely in each text segment).

In the first modality, BB-kb, entity mentions were given and participating systems perform categorization and event extraction. In the second modality, BB-kb+ner, systems had to perform *Bacteria* and *Habitat* entity detection and categorization as well as event extraction.

3 Corpus Construction

3.1 Corpus Selection

The BB corpus consists of titles and abstracts of PubMed entries. We followed a four step procedure to build a representative reference corpus for the task from the whole PubMed database. It started from the set of all PubMed references and successively selected a subset of references while preserving the distribution of bacteria and habitat categories.

In the first step, we selected PubMed entries relevant to bacteria, relying on the MeSH index provided by the NLM. We selected all entries that were indexed by any term of the Organisms/Bacteria subtree (B03). PubMed contained 27,872,481 entries, of which 1,156,824 indexed by a term in the Bacteria subtree (4%).

In the second step, we automatically annotated *Bacteria*, *Geographical* and *Habitat* entities in the title and abstract of these selected entries (see the corpus annotation subsection for details about this automatic approach). We found 6.8 million habitat occurrences, 3.7 million occurrences of bacteria taxon names, and 374 thousand geographical names. This gave us a broad idea of the quantity and diversity of the entries in terms of bacterial taxa and bacterial habitats.

However this collection is too large to be manageable by human annotators. Therefore in the third step, we built a sub-collection of 1,000 entries. We selected the most representative in 2,000 random samples of 1,000 entries. The representativeness was evaluated by the mean squared error (MSE) between the sample and the original collection. We selected the sample with the lowest MSE. The observations from which we computed the MSE included the number of words, the number of occurrences of taxon names for each bacterial family and the number of occurrences of habitat mentions for each top-level concept of OntoBiotope. As expected from a PubMed sample, the majority of entries were biomedical studies. Even though habitats related to human health and welfare are important, the sample does not convey the full diversity of bacteria habitats.

In the fourth step we manually annotated the title and abstract of references from the sample (see section 3.2). As it would require too much human resources the manual annotation of 1,000 PubMed entries is not an option. We randomly picked entries as we finished annotating the previous ones in order to preserve the distribution. The random selection used the same method as the sampling, however we deliberately biased against clinical habitats in order to leave room for more diverse and less frequent habitats.

3.2 Corpus Annotation

Manual annotation was performed by seven annotators with diverse backgrounds: biology, computer science, linguistics, and bioinformatics. Three annotators had annotated documents in the previous editions of the BB task. Each document was annotated by two annotators in a double-blind manner and an adjudication phase resolved disagreements. Annotators relied on detailed guidelines which were revised and clarified when questions arose during the annotation process. The guideline document is available on the BB task website[8].

Annotators used the AlvisAE annotation editor (Papazian et al., 2012). In order to speed up the annotation process, we used Alvis Suite (Ba and Bossy, 2016) to automatically pre-annotate the corpus. It included the Stanford NER tool (Finkel et al., 2005) to annotate *Geographical* locations and the ToMap method (Golik et al., 2011) to detect and categorize *Habitat* entities. *Bacteria* entity automatic recognition and categorization were performed with a rule-based approach relying on a customized dictionary of taxon names, *i.e.* NCBI taxonomy names augmented with typographical variations. Events were extracted using manually defined trigger words and rules in a similar way as Ratkovic et al. (2012). Table 1 gives pre-annotation performance for habitat and bacteria recognition and categorization (cat+ner) and for entity and *Lives_In* event extraction (event+ner). Performance is low, especially for event extraction, which calls into question the benefit of using automatic pre-annotation for these tasks. The low performance of pre-annotation compared to the final gold standard is also an indication that text pre-annotation did not much bias manual an-

---

[8]http://2016.bionlp-st.org/tasks/bb2

notation, since the annotators did not hesitate to make extensive changes in the pre-annotation. We computed the inter-annotator agreement by comparing the individual manual annotations with the consensus, using the same evaluation framework as for the evaluation of participant systems (Section 5). We did not compute any Kappa statistics, since this type of metric is not well-suited for the annotation of textual entities (Hripcsak and Rothschild, 2005). Moreover, even in the case of event annotation, computing Kappa would have been difficult, because event annotation is based on entity annotation. Table 2 shows the agreement of entity boundaries and categorization computed with BB-cat+ner scores and the agreement of entity boundaries and *Lives_In* events computed with BB-event+ner scores. The high precision demonstrates that there was not much disagreement among annotators on the entity boundaries and categorization, or in the *Lives_In* events. The consensus consisted mostly in annotation merging. The lower recall stresses the necessity of multiple annotators in order to ensure that the reference is complete.

	SER	Recall	Precision	F1
cat+ner	1.167	0.287	0.341	0.312
event+ner	1.749	0.187	0.158	0.171

Table 1: Pre-annotation performance (SER = Standard Error Rate)

	Recall	Precision	F1
Entity recog.	0.621	0.955	0.753
Lives_In event	0.311	0.952	0.468

Table 2: Inter-annotator agreement

3.3 Corpus Statistics

Tables 3, 4 and 5 provide descriptive statistics of the corpus for the three subtasks respectively.

They show the distributions of entities, categories, and events among the different datasets (training, development and test) of each subtask. We analyzed these statistics in order to study the characteristics of the BB corpora with respect to the tasks. Each distinct entity surface form has only two occurrences on average in the corpus, which makes the recognition task more difficult than with highly repeated mentions: there are 1,489 and 1,466 distinct entity mentions (*i.e.*, strings or surface forms) out of a total 2,887 and 2,842 annotated entity mentions in the BB-cat and BB-cat+ner datasets, respectively (see Table 3). In comparison, there is less variety in entity categories, since the number of distinct categories is only 519 out of a total of 3,189 occurrences. The combination of these two observations indicates that there is quite a lot of variation in the surface forms of entities, *i.e.*, the same category can be expressed in several different ways in the text. This is particularly true for *Habitat* entities for which there is a higher proportion of distinct surface forms than for *Bacteria* names (59% *vs.* 38% in the combined BB-cat datasets).

Additionally, we computed the proportion of direct mappings (*i.e.*, exact string matches) between *Habitat* surface forms from the training and development datasets of BB-cat and BB-cat+ner and the ontology labels. We found that respectively 24% and 27% Habitat entity occurrences exactly matched with an ontology label. As expected, proportions were similar in the test sets of these two tasks, with respectively 25% and 27% exact matches. This finding emphasizes the fact that there is much variation in the expression of *Habitat* entities, and thus simple methods based on exact string matching are not sufficient to automatically categorize entities with high quality.

Multiple categories may be assigned to a given entity mention, as can be seen in Table 3, which is more challenging than single categories. This is the case mainly for *Habitat* entities, since there is a total of 1,921 distinct *Habitat* entities for a total of 2,221 assigned *Habitat* categories in the BB-cat datasets.

The number of *Geographical* entities in the BB-event+ner sets is much lower than the other entity types with 101 *Geographical* entities only in total, which may make machine-learning approaches less efficient for this type of entity.

Not surprisingly, the majority of *Lives_in* events links *Bacteria* entities to *Habitat* entities and only a small number of events involves *Geographical* entities in the BB-event datasets (*e.g.*, 98 out of a total of 890 events (11%)).

Table 4 also shows the number of intra-sentence *vs.* inter-sentence events, *i.e.* events that involve entity arguments occurring in the same sentence *vs.* events that involve entities occurring in different sentences. The proportion of inter-sentence events is still significant (27%). Methods restricted to the extraction of sentence-level events would suffer from a serious disadvantage.

However the extraction of inter-sentence events is a major challenge, since they are notably more difficult to predict and may require co-reference resolution.

Table 5 details statistics for the knowledge base extraction subtask (BB-kb and BB-kb+ner). Its goal is to build a knowledge base composed of all distinct pairs of Bacteria and Habitat categories linked through the Lives_in relation that can be extracted from the corpus. The number of linked pairs of distinct categories is high with respect to the total number of pairs. There are 185 distinct events out of a total of 312 events in the test set of the BB-kb task (last row of Table 5). This reflects the richness of the information content of the corpus.

4 Shared Task Organization

The BB task schedule was divided in a training period of two months and a test period of twelve days. After the test, detailed evaluation of the system performances was provided to the participating teams and published on the BioNLP-ST 2016 website.

Supporting resources were made available to the participants. These resources are the output of state-of-the-art automated corpus analysis tools applied to the BB datasets. They were generated in the same way as for the SeeDev task of BioNLP-ST (see Chaix et al. (2016) for further details). In addition to the information available on the website, we maintained a set of community web tools. They included a dedicated forum that allowed participants to interact directly with each other and with the organizers, and an online evaluation service the participants could use to evaluate their predictions during the training phase. This service also keeps track of multiple runs allowing participants to monitor their experiments and to compare their predictions to other participant predictions in an anonymous way.

5 Evaluation

The metrics used to evaluate systems depend on the subtasks. When possible we reused metrics from the previous editions so that the results remain comparable.

5.1 BB-cat and BB-cat+ner

BB-cat. For each entity the metrics measures the similarity between the reference category and the predicted category. The overall score is equal to the mean of the similarities for all entities. For *Bacteria* entities the similarity is defined as follows, if the predicted taxon identifier is identical to the reference taxon identifier, then it is set to 1, otherwise 0. For *Habitat* entities we used the same similarity measure as for the 2013 edition of the BB task (Bossy et al., 2013): it is the semantic similarity defined by Wang et al. (2007) with the weight parameter set to 0.65.

BB-cat+ner. The BB subtask was evaluated using the Slot Error Rate (SER), the same method as BioNLP-ST 2013 BB task 1 (Bossy et al., 2013) since the two tasks are the same.

5.2 BB-event and BB-event+ner

The metrics for the evaluation of the BB-event and BB-event+ner subtasks are recall, precision and F-score as for BioNLP-ST 2013 BB task 2 and 3 for the same reasons (Bossy et al., 2013).

5.3 BB-kb and BB-kb+ner

The evaluation of BB-kb submissions is based on the comparison of the reference knowledge base to the one that each participant system has built. The knowledge base associates bacterial taxa with habitat categories. The taxon-habitat category associations are obtained from text-bound *Lives_In* event arguments assigned to taxa and habitat categories. Duplicate associations are removed to generate the knowledge base so that a single association remains between a given taxon and a given habitat category. We applied this procedure to the set of reference events and categories to generate the reference knowledge base and to the events and categories predicted by the participant systems in the same way.

The goal of the BB-kb is to assess how much knowledge a system can extract from a collection of documents. The measure of the exact match between the predicted knowledge base and the reference knowledge base would be too strict and would not satisfy this goal. Thus we designed a measure that evaluate the *similarity* between the two knowledge bases

Each predicted association is paired to the closest reference association using the similarity functions of BB-cat. This process results in each reference association paired to zero (false negative), one, or several predicted associations. Then we can measure the accuracy by which each reference association was found. If the association is not paired to any prediction, then its accuracy is zero, otherwise the accuracy is the mean of the similarity to each prediction. The submissions are evaluated by the mean accuracy for each reference association (mean references). The "mean references" score computes how much the predicted knowledge base maps into the reference knowledge base.

Since the evaluation does not rely on text-bound annotations, the BB-kb+ner was evaluated with the same metrics as BB-kb.

6 Results

A total of 14 teams participated in Bacteria Biotope 2016. They were from several countries: Turkey (BOUN), France (LIMSI), Denmark (TagIt), Canada (VERSE), Finland (TurkuNLP, UTS, UMS) and China (DUTIR, WhuNlpRE, HK, whunlp, WXU). Two participants retracted their submissions (they correspond to blank lines in result tables). We present the results obtained by the participating teams. Detailed results are available on the task page[9].

6.1 Performance on BB-cat / BB-cat+ner

The results of systems that participated to BB-cat (2 teams) and BB-cat+ner subtasks (3 teams) are given in Table 6 and 7, respectively.

Team	Prec. all	Prec. Bacteria	Prec. Habitat	Prec. Multi cat.
BOUN	**0.679**	**0.801**	**0.620**	0.486
LIMSI	0.503	0.637	0.438	**0.516**

Table 6: Team results for the BB-cat task ("Prec." = "Precision"; "Multi cat." = "Multiple categorizations")

BOUN achieved the best performance for the categorization task (BB-cat) with 0.679 precision. As expected, performance was much higher for the categorization of Bacteria entities (0.801 for the best precision) than for that of Habitat entities (0.62). Bacteria are usually referred to using names from the NCBI taxonomy with a few variations, while Habitats are mainly noun and adjectival phrases that are expressed in many ways and may be very different from their concept label form. Moreover, Habitats may be categorized using several ontology concepts, which creates an additional difficulty. The last column of Table 6 shows results for multiple categorization

cases. The LIMSI team obtained stable performance while the BOUN team performed significantly lower than for all entities (0.486 *vs.* 0.679).

When taking into account entity recognition in addition to categorization (BB-cat+ner, Table 7), TagIt achieved the best SER (0.628), and the difference between the top and last teams is significant (0.27 points). As for the BB-cat task, systems performed better on Bacteria entities than on Habitat entities. We also assessed the performance of entity recognition (without taking into account categorization), *i.e.*, systems are evaluated for their ability to predict entity boundaries in the text (see the bottom part of Table 7). The results of boundary detection also reflect the difference in difficulty between *Habitat* and *Bacteria* entities.

Compared to the Bacteria Biotope 2013 edition, the performance seems to have dropped. The best SER for *Habitat* entity recognition and categorization was 0.661 (Bossy et al., 2015), while it is 0.775 this year. This may be due to the change of document source, *i.e.*, scientific dense documents instead of general purpose web pages. It may also be due to the higher proportion of cases of multiple category assignments, while these cases remained marginal in the 2013 edition. Another reason may be the high number of clinical studies where the distinction between categories (*e.g.*, treated and non-treated patients, pediatric and adult patients) may require a more thorough analysis of the event context. Therefore the task also entails co-reference resolution.

		TagIt	LIMSI	whunlp
Overall	SER	**0.628**	0.827	0.901
	Recall	**0.456**	0.361	0.273
	Precision	**0.612**	0.486	0.407
Bacteria	SER	**0.399**	0.771	0.823
	Recall	**0.692**	0.539	0.397
	Precision	**0.857**	0.623	0.637
Habitat	SER	**0.775**	0.862	0.950
	Recall	**0.303**	0.246	0.193
	Precision	**0.430**	0.371	0.275
Bacteria boundaries	SER	**0.236**	0.277	0.436
	Recall	**0.772**	0.751	0.565
	Precision	**0.954**	0.903	0.893
Habitat boundaries	SER	0.599	**0.597**	0.627
	Recall	0.476	**0.504**	0.493
	Precision	0.675	**0.728**	0.690

Table 7: Team results for the BB-cat+ner task

6.2 Performance on BB-event / BB-event+ner

Among subtasks, the event extraction subtask (more specifically the BB-event task) attracted the most participants, with a total of eleven differ-

[9]http://2016.bionlp-st.org/tasks/bb2/bb3-evaluation

ent teams, three of which participated in the BB-event+ner subtask and eleven in the BB-event subtask. Tables 8 and 9 show team performances on BB-event and BB-event+ner tasks respectively.

VERSE obtained the highest F1 score for the BB-event task (0.558). The difference between the top and last teams is only 0.10 points and participants ranked 4th to 11th obtained very similar results (ranging from 0.474 to 0.455 F1 score). All participants achieved better performance when predicting *Lives_in* events with *Geographical* arguments than events with *Habitat* arguments (5th and 6th columns of Table 8), although events with *Geographical* arguments are less frequent. The reason could be that most of *Geographical* entities are linked to a *Bacteria* entity, which makes the decision easier than for *Habitat* entities, for which there are many occurrences that are not involved in any *Lives_in* event.

Not surprisingly systems had less trouble predicting intra-sentence events than inter-sentence events, as all yielded significantly higher F1 score on intra-sentence events (see last column of Table 8). Detailed analysis of the predictions made by the systems shows that LIMSI was the only team to consistently predict inter-sentence events. Other systems predicted roughly the same number of events when considering only intra-sentence events or all events together in the evaluation.

There is a drastic drop in performance when adding entity recognition to the event extraction task (BB-event+ner task, see Table 9). All three participating teams obtained very similar results in terms of F1 score, although the balance between precision and recall differs. The LIMSI team (ranked 1st) achieved a perfect balance between precision and recall, while UTS and the WhuNlpRE team obtained much higher precision but lower recall. As for the BB-event task, performances are significantly higher for *Lives_in* events involving *Geographical* entities, and intra-sentence events.

For both tasks, systems performed better in average than in the 2013 edition. Indeed, the best F1 scores (Bossy et al., 2015) were 0.49 for the detection of localization events (*vs.* 0.558 for *Lives_in* events in this edition) and 0.14 for the combination of entity recognition and event extraction (vs. 0.19). This suggests that participant methods have improved and become more accurate. However, the F1-score for BB-event+ner remains relatively low, which directly results from the combined complexity of the two sub-problems in the same task.

6.3 Performance on BB-kb / BB-kb+ner

Only the LIMSI team participated in the knowledge base extraction subtask. Results are given in Table 10 for both the BB-kb and BB-kb+ner tasks. The LIMSI system for BB-cat (Table 6) and BB-event (Table 8) provides a good reconstruction of the knowledge base (BB-kb) which highlights the fact that automatic categorization and event extraction methods are already efficient for the task of knowledge base construction. However, the performance is significantly lower when reference entities are not provided. This large gap in performance may be explained by the difficulty of recognizing entities (as also shown in the BB-cat+ner task), and the fact that a fair amount of entities is not repeated in the corpus. Consequently the false negatives in entity detection have a strong impact on the end-to-end task of knowledge base construction.

	LIMSI	UTS	WhuNlpRE
F1	**0.192**	0.190	0.182
Recall	**0.191**	0.133	0.111
Precision	0.193	0.331	**0.498**
F1 (Habitat)	0.186	0.174	**0.196**
F1 (Geographical)	0.283	**0.350**	NA
F1 (Intra-sentence)	**0.286**	0.234	0.232

Table 9: Team results for the BB-event+ner task

	BB-kb	BB-kb+ner
LIMSI	0.771	0.202

Table 10: Results for BB-kb and BB-kb+ner (mean-references measure)

6.4 Systems

Systems used different resources and methods depending on the sub-tasks.

Entity Detection and Categorization. Systems used dictionary-based (TagIt) and machine-learning based (LIMSI, WhuNlpRE, UTS) methods to detect entity mentions in text in the BB-cat+ner and BB-event+ner subtasks. All relied on existing terminology and ontology resources, including the NCBI Taxonomy, the List of Prokaryotic Names with Standing in Nomenclature (Parte, 2013), the Brenda Tissue Ontology (Gremse et al., 2011), the Environment Ontology (Buttigieg et al.,

2013), the OntoBiotope ontology, and WordNet (Fellbaum, 1998). The TagIt system performed dictionary matching coupled with acronym detection and heuristic rules to adjust entity boundaries. The LIMSI team used conditional random fields (CRFs) and the WhuNlpRE team used neural networks. Both these teams generated rich features for their machine-learning algorithms: lexical, morpho-syntactic, dictionary projection, existing named entity recognition tools, Brown clustering, and word embeddings. The UTS team relied on Support Vector Machines (SVM) with features based on the output of existing NER tools provided by the organizers as supporting resources. The rule-based approach of the TagIt system achieved the highest performance in entity detection and categorization (BB-cat+ner), although the CRF approach of the LIMSI system was the most accurate in Habitat boundary detection.

Teams relied on rule-based (TagIt, LIMSI) and similarity-based (BOUN) approaches to categorize entities in the BB-cat and BB-cat+ner subtasks. The TagIt system performed entity categorization jointly with entity detection using dictionaries and rules. The BOUN team combined approximate string matching (edit distance) with an Information Retrieval based bag-of-word approach (cosine similarity of word vectors weighted with the tf-idf). This approach was the most successful in the BB-cat.

Prediction of Events. All systems used machine-learning to predict *Lives_in* events. The most popular algorithms are SVM (VERSE, HK, UTS, LIMSI) and neural networks (TurkuNLP, WhuNlpRE, DUTIR). UMS combined predictions from a SVM and a neural network. Most systems rely on syntactic parsing to generate features (VERSE, TurkuNLP, UMS, HK, DUTIR, UTS). Other common features included part-of-speech tags, word embeddings (trained on large corpora, *e.g.*, large sets of PubMed abstracts), and entity recognition. Rankings do not show any correlation to the machine learning algorithm, for instance the top ranking is based on SVM and the second is based on neural networks. Therefore, no conclusion can be drawn on the most appropriate class of methods. The quality of the predictions seems to rely mainly on the feature design, *i.e.*, what types of feature were used by the systems. To this respect the two top ranking systems have syntactic parsing-based features. More specifically, they both generate features based on the dependency path between entities.

7 Conclusion

The interest for the Bacteria Biotope Task keeps growing with a total of 14 teams participating in this third edition, and showing very promising results. 11 teams participated in the event extraction task (BB-event), demonstrating the interest of the NLP community for this challenging subject. For this event detection task, the most commonly used methods were SVMs and neural networks, and they yielded higher performance than during the 2013 edition of the task. However, a detailed analysis of the results showed that inter-sentence events still remain a challenge and are ignored by most systems. The other BB tasks, *i.e.* entity detection and categorization and knowledge base extraction, attracted fewer participants in comparison to event extraction. Knowledge base population was the most challenging task, since it required a large range of skills.

To help participants, supporting resources were provided but they were not much used. A more thorough investigation is needed to better understand the needs of participants in terms of external resources. The introduction of the online evaluation service with detailed metrics appears to have facilitated the development cycle of predictive systems. This service will be maintained online allowing for future experiments and comparisons with BB'16 data.

	BB-cat				**BB-cat+ner**			
	Train	Dev	Test	Total	Train	Dev	Test	Total
Documents	71	36	54	*161*	71	36	54	*161*
Words	16,295	8,890	13,797	*38,982*	16,295	8,890	13,933	*39,118*
Bacteria	375	244	347	*966*	375	244	401	*1,020*
Habitat	747	454	720	*1,921*	747	454	621	*1,822*
Total entities	*1,122*	*698*	*1,067*	*2,887*	*1,122*	*698*	*1,022*	*2,842*
Distinct *Bacteria*	167	111	146	*364*	167	111	181	*393*
Distinct *Habitat*	476	267	478	*1,125*	476	267	416	*1,073*
Total distinct entities	*643*	*378*	*624*	*1,489*	*643*	*378*	*597*	*1,466*
Bacteria categories	376	245	347	*968*	376	245	401	*1,022*
Habitat categories	825	535	861	*2,221*	825	535	681	*2,041*
Total categories	*1,201*	*780*	*1,208*	*3,189*	*1,201*	*780*	*1,082*	*3,063*
Distinct *Bacteria* categories	85	70	80	*190*	85	70	87	*193*
Distinct *Habitat* categories	210	122	177	*329*	210	122	168	*341*
Total distinct categories	*295*	*192*	*257*	*519*	*295*	*192*	*255*	*534*

Table 3: Descriptive statistics of the corpus for BB-cat and BB-cat+ner

	BB-event				**BB-event+ner**			
	Train	Dev	Test	Total	Train	Dev	Test	Total
Documents	61	34	51	*146*	71	36	54	*161*
Words	13,850	8,491	13,039	*35,380*	16,295	8,890	13,933	*39,118*
Bacteria	358	238	336	*932*	375	244	401	*1,020*
Habitat	687	454	720	*1,861*	747	454	621	*1,822*
Geographical	35	38	37	*110*	36	38	27	*101*
Total entities	*1,080*	*730*	*1,093*	*2,903*	*1,158*	*736*	*1,049*	*2,943*
Lives_in events (*Habitat*)	294	186	312	*792*	294	186	288	*768*
Lives_in events (*Geog.*)	33	37	28	*98*	33	37	26	*96*
Intra-sentence events	240	165	248	*653*	240	165	231	*636*
Inter-sentence events	87	58	92	*237*	87	58	83	*228*
Total Lives_in events	*327*	*223*	*340*	*890*	*327*	*223*	*314*	*864*

Table 4: Descriptive statistics of the corpus for BB-event and BB-event+ner

	BB-kb				**BB-kb+ner**			
	Train	Dev	Test	Total	Train	Dev	Test	Total
Documents	61	34	50	*145*	71	36	54	*161*
Words	13,850	8,491	12,758	*35,099*	16,295	8,890	13,933	*39,118*
Bacteria	358	238	330	*926*	375	244	401	*1,020*
Habitat	687	454	720	*1,861*	747	454	621	*1,822*
Total entities	*1,045*	*692*	*1,050*	*2,787*	*1,122*	*698*	*1,022*	*2,842*
Bacteria categories	359	239	330	*928*	376	245	401	*1,022*
Habitat categories	765	535	861	*2,161*	825	535	681	*2,041*
Total categories	*1,124*	*774*	*1,191*	*3,089*	*1,201*	*780*	*1,082*	*3,063*
Distinct *Bacteria* categories	81	69	77	*183*	85	70	87	*193*
Distinct *Habitat* categories	197	122	177	*317*	210	122	168	*341*
Total distinct categories	*278*	*191*	*254*	*500*	*295*	*192*	*255*	*534*
Lives_in events	294	186	312	*792*	294	186	288	*768*
Distinct Lives_in events	*204*	*156*	*185*	*522*	*204*	*156*	*183*	*524*

Table 5: Descriptive statistics of the corpus for BB-kb and BB-kb+ner

Team	F1	Recall	Precision	F1 (*Habitat*)	F1 (*Geo.*)	F1 (Intra-sentence)
VERSE	**0.558**	0.615	0.510	**0.545**	0.714	0.634
TurkuNLP	0.521	0.448	**0.623**	0.499	**0.755**	0.620
LIMSI	0.485	**0.646**	0.388	0.482	0.525	**0.636**
HK	0.474	0.392	0.599	0.452	0.708	0.567
WhuNlpRE	0.471	0.407	0.559	0.471	0.465	0.561
UMS	0.463	0.399	0.551	0.439	0.704	0.550
DUTIR	0.456	0.382	0.566	0.451	0.512	0.544
WXU	0.455	0.383	0.560	0.445	0.578	0.540
-	-	-	-	-	-	-
UTS	0.451	0.382	0.551	0.425	0.704	0.537
-	-	-	-	-	-	-

Table 8: Team results for the BB-event task

References

Mouhamadou Ba and Robert Bossy. 2016. Interoperability of corpus processing work-flow engines: the case of alvisnlp/ml in openminted. In Richard Eckart de Castilho, Sophia Ananiadou, Thomas Margoni, Wim Peters, and Stelios Piperidis, editors, *Proceedings of the Workshop on Cross-Platform Text Mining and Natural Language Processing Interoperability (INTEROP 2016) at LREC 2016*, pages 15–18, Portoroz, Slovenia, May. European Language Resources Association (ELRA).

Robert Bossy, Julien Jourde, Alain-Pierre Manine, Philippe Veber, Erick Alphonse, Maarten Van De Guchte, Philippe Bessières, and Claire Nédellec. 2012. Bionlp shared task-the bacteria track. *BMC bioinformatics*, 13(11):1.

Robert Bossy, Wiktoria Golik, Zorana Ratkovic, Philippe Bessières, and Claire Nédellec. 2013. Bionlp shared task 2013–an overview of the bacteria biotope task. In *Proceedings of the BioNLP Shared Task 2013 Workshop*, pages 161–169.

Robert Bossy, Wiktoria Golik, Zorana Ratkovic, Dialekti Valsamou, Philippe Bessières, and Claire Nédellec. 2015. Overview of the gene regulation network and the bacteria biotope tasks in bionlp'13 shared task. *BMC bioinformatics*, 16(10):1.

Pier Luigi Buttigieg, Norman Morrison, Barry Smith, Christopher J Mungall, and Suzanna E Lewis. 2013. The environment ontology: contextualising biological and biomedical entities. *Journal of biomedical semantics*, 4(1):1.

Estelle Chaix, Bertrand Dubreucq, Abdelhak Fatihi, Dialekti Valsamou, Robert Bossy, Mouhamadou Ba, Louise Deléger, Pierre Zweigenbaum, Philippe Bessières, Loic Lepiniec, and Claire Nédellec. 2016. Overview of the regulatory network of plant seed development (seedev) task at the bionlp shared task 2016. In *Proceedings of the 4th BioNLP Shared Task workshop*, Berlin, Germany, August. Association for Computational Linguistics.

Scott Federhen. 2012. The ncbi taxonomy database. *Nucleic acids research*, 40(D1):D136–D143.

C Fellbaum. 1998. Wordnet: An on-line lexical database and some of its applications.

Jenny Rose Finkel, Trond Grenager, and Christopher Manning. 2005. Incorporating non-local information into information extraction systems by gibbs sampling. In *Proceedings of the 43rd Annual Meeting on Association for Computational Linguistics*, pages 363–370. Association for Computational Linguistics.

Melissa Merrill Floyd, Jane Tang, Matthew Kane, and David Emerson. 2005. Captured diversity in a culture collection: case study of the geographic and habitat distributions of environmental isolates held at the american type culture collection. *Applied and Environmental Microbiology*, 71(6):2813–2823.

Wiktoria Golik, Pierre Warnier, and Claire Nédellec. 2011. Corpus-based extension of termino-ontology by linguistic analysis: a use case in biomedical event extraction. In *WS 2 Workshop Extended Abstracts, 9th International Conference on Terminology and Artificial Intelligence*, pages 37–39.

Marion Gremse, Antje Chang, Ida Schomburg, Andreas Grote, Maurice Scheer, Christian Ebeling, and Dietmar Schomburg. 2011. The brenda tissue ontology (bto): the first all-integrating ontology of all organisms for enzyme sources. *Nucleic acids research*, 39(suppl 1):D507–D513.

George Hripcsak and Adam S Rothschild. 2005. Agreement, the f-measure, and reliability in information retrieval. *Journal of the American Medical Informatics Association*, 12(3):296–298.

Natalia Ivanova, Susannah G Tringe, Konstantinos Liolios, Wen-Tso Liu, Norman Morrison, Philip Hugenholtz, and Nikos C Kyrpides. 2010. A call for standardized classification of metagenome projects. *Environmental microbiology*, 12(7):1803–1805.

Jin-Dong Kim, Tomoko Ohta, Sampo Pyysalo, Yoshinobu Kano, and Jun'ichi Tsujii. 2009. Overview of bionlp'09 shared task on event extraction. In *Proceedings of the Workshop on Current Trends in Biomedical Natural Language Processing: Shared Task*, pages 1–9. Association for Computational Linguistics.

Jin-Dong Kim, Sampo Pyysalo, Tomoko Ohta, Robert Bossy, Ngan Nguyen, and Jun'ichi Tsujii. 2011. Overview of bionlp shared task 2011. In *Proceedings of the BioNLP Shared Task 2011 Workshop*, pages 1–6. Association for Computational Linguistics.

Claire Nédellec, Robert Bossy, Jin-Dong Kim, Jung-Jae Kim, Tomoko Ohta, Sampo Pyysalo, and Pierre Zweigenbaum. 2013. Overview of bionlp shared task 2013. In *Proceedings of the BioNLP Shared Task 2013 Workshop*, pages 1–7.

Frédéric Papazian, Robert Bossy, and Claire Nédellec. 2012. Alvisae: a collaborative web text annotation editor for knowledge acquisition. In *Proceedings of the Sixth Linguistic Annotation Workshop*, pages 149–152. Association for Computational Linguistics.

Aidan C Parte. 2013. Lpsnlist of prokaryotic names with standing in nomenclature. *Nucleic acids research*, page gkt1111.

Miguel Pignatelli, Andrés Moya, and Javier Tamames. 2009. Envdb, a database for describing the environmental distribution of prokaryotic taxa. *Environmental Microbiology Reports*, 1(3):191–197.

Zorana Ratkovic, Wiktoria Golik, and Pierre Warnier. 2012. Event extraction of bacteria biotopes: a knowledge-intensive nlp-based approach. *BMC bioinformatics*, 13(11):1.

James Z Wang, Zhidian Du, Rapeeporn Payattakool, S Yu Philip, and Chin-Fu Chen. 2007. A new method to measure the semantic similarity of go terms. *Bioinformatics*, 23(10):1274–1281.

Refactoring the Genia Event Extraction Shared Task
Toward a General Framework for IE-Driven KB Development

Jin-Dong Kim[*,1]**, Yue Wang**[1]**, Nicola Colic**[2]**, Seung Han Baek**[3]**, Yong Hwan Kim**[3]**, Min Song**[3]
[1] Database Center for Life Science (DBCLS)
[2] University of Zürich
[3] Yonsei University

Abstract

For its fourth organization, the Genia event extraction (GE) shared task is refactored toward a general platform for shared information extraction (IE) tasks, and for an IE-driven knowledge base (KB) system. On the newly implemented shared task platform, the GE task is run as an experimental task. The task and the platform has been tested by two teams who cooperated with the organizers. The paper presents the new shared task system and discusses on the experimental submissions.

1 Introduction

Since its first introduction in 2009 as the task of the first BioNLP Shared Task (BioNLP-ST) organization, the Genia event extraction (GE) task has been one of the most investigated IE tasks (Kim et al., 2009; Kim et al., 2011; Kim et al., 2013). The biggest contribution of BioNLP-ST might be that it introduced fine-grained and highly structured information extraction (IE) tasks to the community of biomedical information extraction (BioIE), when the research in the community was weighted toward extracting binary relations (Krallinger et al., 2007; Lu et al., 2004; Chun et al., 2006). Since then the tasks of BioNLP-ST have motivated and nourished the community to develop a number of biomedical event extraction systems (Björne and Salakoski, 2013; Miwa et al., 2010),

Originally designed as tasks based on intrinsic evaluation, however, the tasks of BioNLP-ST could not be free from criticism on unclarity about their impact on real world application (Caporaso et al., 2008). Also, there was a growing need for generalized resources for shared task organization with which the cost of organizing shared tasks

could be substantially reduced. With this motivation, for its 4th organization in 2016, the GE task is completely re-designed and re-implemented as an experimental task with two goals.

Firstly, we aim at establishing a seamless connection from the IE task to knowledge base (KB) construction. It means we assume KB construction as the target application of the GE task. Particularly, we aim at developing a KB about NFκB proteins, which is the subject domain the GE task has focused on. In the end, we hope to be able to deliver an end-user service of the KB, so that people who are interested in NFκB proteins can easily access knowledge about them. Toward this end, we automate the process of populating a KB from the output of the task, and solicit working systems to perform the task.

Secondly, we aim at generalizing the resources of shared task organization. Previous iterations of organization showed that shared task is an effective format to promote development of IE solutions. Shared task organization however requires a lot of effort and expertise. If the resources for shared task organization become generalized and readily available, more shared tasks can be easily organized. To this end, we re-designed and re-implemented the shared task resources which have been developed so far for the GE task.

Due to the complexity of refactoring the whole task, instead of being run as a competition among participants, the GE4 task is organized as an experimental task, experimenting newly implemented features, with involvement of voluntary feedback from participants. Finally, two systems could go through up to their final submissions, thanks to which the newly implemented shared task system could be thoroughly tested. Manual analysis on the submissions shows both achievments and remaining issues, which are discussed in the end of this paper.

---

*Corresponding author, `jdkim@dbcls.rois.ac.jp`

Proceedings of the 4th BioNLP Shared Task Workshop, pages 23–31,
Berlin, Germany, August 13, 2016. ©2016 Association for Computational Linguistics

2 Design

2.1 Platform

To achieve the first goal of generalizing the shared task system, *PubAnnotation* (Kim and Wang, 2012) was chosen as the platform. There were several reasons for the choice. Firstly, as a public repository of literature annotation, PubAnnotation provides various ways of submitting and accessing annotation data sets, which are fundamental for shared task organization. Secondly, it features an automatic text alignment function, which provides a reliable solution for aligning annotations collected from different groups. Thirdly, it is a near mature system, which has a growing user base with more than hundred of data sets.

While PubAnnotation provides many useful functions, a shared task organization still requires more functions. Most importantly, automatic evaluation needs to be enabled for efficient development of IE systems. Also, to prevent over-fitting the benchmark data set, often the annotations in the benchmark data set are required to be hidden. Accordingly, the two key features are implemented into PubAnnotation, which are described in following sections.

2.1.1 Comparison of annotations

A shared task organization often features an automatic evaluation of predicted annotations. For generalization, we cast it as general comparison of two different annotation sets. On PubAnnotation, an annotation data set is maintained as a *project*, and each project is maintained by its *maintainer*.

A new feature *annotation comparison* is implemented into PubAnnotation. Using the feature, the maintainer of a project can compare the project against any other project. We call the former a *subject project*, and the latter a *reference project*. A comparison is performed by looking at how many annotations in the reference project can be recovered in the subject project. The comparison is calculated in terms of recall, precision, and f-score, in their standard meaning.

As PubAnnotation represents annotations in three types, *denotations*, *relations*, and *modifications*[1], comparison is also performed for each of the three types. In case the subject and reference projects have different sets of documents, comparison is performed only for the documents found in

[1]http://www.pubannotation.org/docs/
annotation-format/

both projects.

With this feature, any corpus with manual annotation can potentially serve as a shared task: any one can attempt to automatically reproduce the manual annotation, and evaluate the accuracy.

2.1.2 Blind annotations

A new feature *blind annotations* is implemented into PubAnnotation, to enable hiding annotations in a certain project. By blinding annotations of a project, individual annotations become inaccessible. However, the project can still be used for comparison. In this way, the project can still function as a benchmark data set.

2.2 Data sets

Data sets prepared for the GE4 task is grouped into *benchmark data sets* and *supporting data sets*.

2.2.1 Benchmark data sets

For the benchmark data set of the GE4 task, the same set of documents used for the GE3 task are cleaned and used again. However, the separation of the data set into training, development and test sets is slightly changed: the training and development data sets are merged into one set which we call a *reference data set*. Thus the GE4 benchmark data set consists of two sets: the reference data set with 20 full papers and the test data set with 14 full papers. The change in dataset separation and naming is made in order to remove the impression that it is a machine learning task and to encourage development of various approaches.

The annotations in the test data set are "blinded" using the newly implemented feature (see section 2.1.2). Following the tradition of BioNLP-ST to provide protein annotations for the test data set, which will allow participants to spend more time for developing their event extraction system, the test data set is duplicated to make what we call a *test-start data set*. The test-start data set is the same as the test data set except for the fact that it has only protein annotations and the annotations are not blinded. Participant can begin their test first by obtaining a copy of the test-start data set. Then, event annotations produced by their systems can be added to it, which will be compared against the test data set for evaluation. The three benchmark data sets for the GE4 task are illustrated on the top of Figure 1.

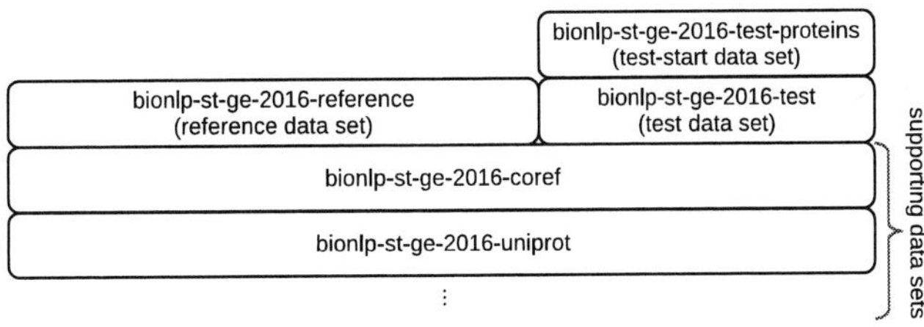

Figure 1: Data sets for the GE4 task

2.2.2 Supporting data sets

Besides the benchmark data sets, other data sets are prepared to support participants to use rich information. Firstly, the coreference annotations from the GE3 data set are separated into an individual annotation set, *bionlp-st-ge-2016-coref*. Secondly, UniProt IDs are annotated to the benchmark data sets, to provide "normalization" or "grounding" of protein annotations. Note that the GE4 task organization aims at constructing an IE-driven KB which requires information pieces to be grounded to database entries. The UniProt ID annotation thus plays an important role in the GE4 task. For the UniProt ID annotation, a simple dictionary matching approach is used, but the dictionary is tailored to the benchmark data sets to raise the accuracy of UniProt ID annotation particularly for the benchmark data sets.

For other supporting data sets, we attempted to collect automatic annotation tools, rather than just collecting static annotation data sets[2]. PubAnnotation has a feature to communicate with web services to obtain annotations, and the feature is used to produce the supporting resources via the automatic annotation tools. It ensures that the same annotations can be produced for new documents. Besides the two sets of annotations described above, two syntactic parsers, and several named entity recognizers are prepared as RESTful web service:

- bionlp-st-ge-2016-uniprot: UniProt ID annotation

- bionlp-st-ge-2016-coref: coreference annotation

- pmc-enju-pas: deep dependency parsing by Enju (Miyao and Tsujii, 2008)

- bionlp-spacy-parsed: dependency parsing spacy (Honnibal et al., 2013)

- UBERON-AE: anatomical entities in UBERON (Mungall et al., 2012)

- ICD10: disease names as defined in ICD10

---

[2]Except for the coreference annotation, which is originally produced manually.

- GO-BP: biological processes as defined in GO

- GO-CC: cellular components as defined in GO

Note that collection of supporting annotations usually requires a non-trivial effort of organizers, to ensure all the annotations provided by different groups to be precisely aligned to the texts in the benchmark datasets. Otherwise, there is a high chance that the texts may be changed during pre-processing by different groups, which may cause an issue of aligning different versions of texts when they are collected. However, thanks to the automatic alignment algorithm implemented in PubAnnotation (See section 2.1), it is not an issue any more as long as they are collected on PubAnnotation. It is a clear benefit of using PubAnnotation as a platform of shared task organization.

Figure 9 shows excerpts of data sets prepared for the GE4 task. The annotation data sets can be retrieved individually or altogether through the RESTful API. For example, by accessing the following URL, the annotations shown in Figure 9 can be obtained in JSON at once:

```
http://pubannotation.org/docs/
sourcedb/PMC/sourceid/3245220/divs/
11/spans/4375-4513/annotations.json?
projects=bionlp-st-ge-2016-reference,
bionlp-st-ge-2016-uniprot,
bionlp-st-ge-2016-coref,pmc-enju-pas,
bionlp-spacy-parsed,GO-BP
```

2.3 KB

By the KB, we mean a SPARQL endpoint populated with RDF statements which are results of conversion from the GE task results. To achieve the goal of establishing a seamless connection from the IE to KB, an automatic process is designed and implemented into PubAnnotation for:

- conversion of annotations to RDF statements, and

- feeding the statements into a SPARQL endpoint.

Also, a SPARQL-driven user interface to search the KB is designed and implemented.

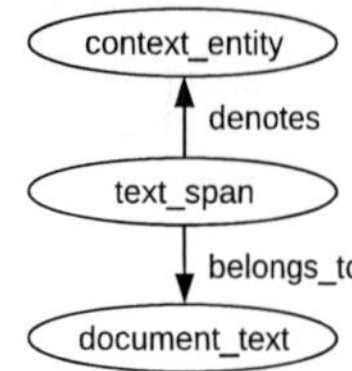

Figure 2: The core model of TAO

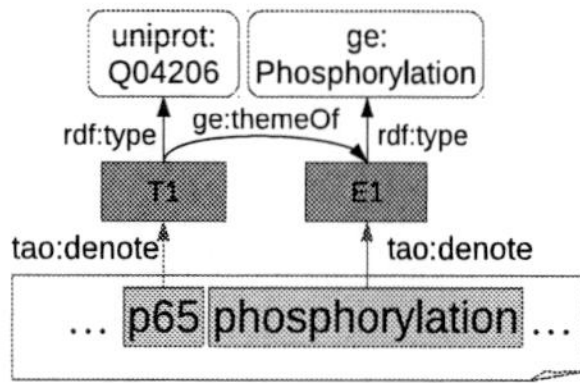

Figure 3: Annotation example using TAO

Considering its characteristics, the KB is designed to provide an easy access to the textual contexts of each knowledge piece. After surveying existing vocabularies for RDF statements (Ciccarese et al., 2011; Livingston et al., 2013), we chose to use a minimal vocabulary optimized for search, which we call *text annotation ontology (TAO)* (Kim et al., 2015). Figure 2 shows the core model of TAO, and Figure 3 shows an example of annotation representation using TAO. The example describes that

- the span *p65* "denotes" *T1*.

- *T1* is a *uniprot:Q04206*.

- the span *phosphorylation* "denotes" *E1*.

- *E1* is a *ge:Phosphorylation*.

- *T1* is a theme of *E1*.

Note that the role of TAO is to make connections between the two text spans, *p65* and *phosphorylation*, and the corresponding context entities, *T1* and *E1*, respectively[3]. Other parts of the annotations are described using other vocabularies: look at the two namespaces, *rdf* and *ge*.

A converter to produce RDF statements from annotations and a loader to feed the statements to a SPARQL endpoint is implemented to create an automatic flow from IE results to KB. TAO makes

---

[3]The prefixes, *T* and *E*, are used here just for readability. They do not hold any special meaning in the system.

SPARQL queries to search the KB simple. For example, following query instructs the system to search for spans (*?s*) that denote an object (*?o*) which is a *uniprot:q04206*.

```
PREFIX tao:<http://pubannotation.org/ontology/tao.owl#>
PREFIX prj:<http://pubannotation.org/projects/>
PREFIX uniprot:<http://www.uniprot.org/uniprot/>

SELECT ?s
FROM prj:bionlp-st-ge-2016-uniprot
WHERE {
  ?s tao:denotes ?o .
  ?o a uniprot:Q04206 .
}
```

The results are URIs of the spans:

```
doc:sourcedb/PMC/sourceid/2664230/divs/2/spans/818-821
doc:sourcedb/PMC/sourceid/2664230/divs/5/spans/1128-1131
doc:sourcedb/PMC/sourceid/2674207/divs/18/spans/2512-2515
...
```

Note, however, that the span URIs are dereferenceable URIs which PubAnnotation provides. This means that the user can directly access the span following the URI. Figure 4 shows the spans of URIs from the above example rendered in TextAE[4], the default visualizer of PubAnnotation.

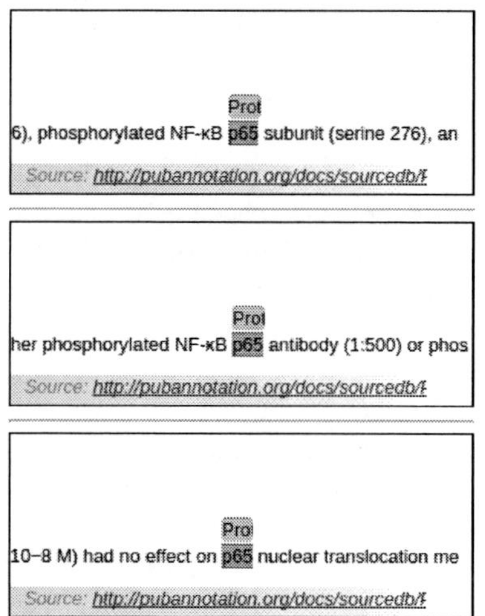

Figure 4: Example of spans rendered in TextAE

2.4 Participation procedure

Participants to the GE4 task are supposed to go through following procedure:

1. To create a new project in PubAnnotation.

2. To import documents from the project, *bionlp-st-2016-test-proteins* to the new project. The 14 documents in the test set will be copied into the new project.

3. To import also annotations from the project, *bionlp-st-2016-test-proteins* to the project. All the protein annotations in the test set will be copied into the project.

---

[4]http://textae.pubannotation.org

4. At this point, the participant may want to compare the project against the test project. It will show that protein annotations are 100% correct, but the other annotations, e.g., events, are of 0%.

5. To produce event annotations, using a participating system, upon the protein annotations.

6. To upload the annotations to the project.

7. To compare the project against to the test project.

Every step of the procedure can be performed using the graphical interface of PubAnnotation. Some steps also can be performed using a programmable RESTful API of PubAnnotation. We believe the procedure is quite generic and can be applied to other shared tasks with similar setting.

3 Results and analyses

The results of GE4 organization are as follows:

- The general shared task framework implemented in PubAnnotation.

- The GE4 task re-engineered using the new framework

- The pipeline to populate a KB (SPARQL endpoint) from IE results

- The user interface to the KB

- The user experience of participants

As the first three are explained in previous sections, this section discusses the last two: KB user interface and user experience. Also, the benchmark data sets are analyzed to simulate the process of knowledge access using the KB.

3.1 User interface to IE-driven KB

A prototype interface to the IE-driven KB is implemented, of which a snapshot is shown in Figure 5. Since the KB is implemented as RDF data sets stored in a SPARQL endpoint, the interface is also SPARQL-oriented: see the input box for a SPARQL query in the center of the interface.

For those who are not familiar with SPARQL, a template system is implemented. A SPARQL template is a SPARQL query with placeholders, of which the value is easily changeable by user's input. For example, look at the template shown in Figure 6. It has one placeholder, __uniprot_id__. A placeholder is indicated by double underscore characters ('__') at its both sides. The title of the template is supposed to have the same placeholders. When displayed, the placeholders in the title become text input boxes to accept user's input,

as shown at the top in the left pane of the screenshot. Upon change of the value in the input boxes, the placeholders in the SPARQL template are also updated, accordingly. Using the templates, users who are not familiar with SPARQL can still access the KB. Even for expert SPARQL users, it reduces time to author frequently necessary queries from scratch. In the left pane of the snapshot, 7 predefined templates are shown.

The next section presents results of analyzing benchmark data sets utilizing the templates.

3.2 Data analysis from KB perspective

In this section, the benchmark data sets are analyzed from a perspective of KB, and observations are discussed.

Table 1 shows statistics of UniProt ID annotations, which form the basis of the knowledge pieces of the KB we develop. For accuracy, only the UniProt ID annotations that are overlapping with (manual) protein annotations are counted. Note that, UniProt ID annotations that are not annotated as proteins in the benchmark data set are not involved in any further annotations, e.g. relations, so, anyway, they cannot be involved in any knowledge piece to be extracted from the data sets.

	Reference	Test	Sum
No. of instances	8,292	3,148	11,440
No. of types	221	110	242

Table 1: Statistics of UniProt ID annotation

Template 1, *Find all the proteins in the benchmark data sets*, with slight modifications, e.g. addition of *GROUP BY* modifier to count types, is used to obtain the statistics. Among the 110 UniProt IDs that appear in the test data set, 21 do not appear in the reference data set, simulating unseen protein names. They may represent an extra challenge for protein name recognition, and an extra chance for novel knowledge piece, at the same time.

Table 2 shows statistics of NFκB proteins, for which Template 3, *Find all the contexts where the protein __uniprot_id__ appears*, is used with the UniProt IDs of the 5 NFκB proteins set to the placeholder. It shows that *p65* is the most frequently referenced protein in both reference and test data sets.

One of typical search needs would be to find the proteins that regulate a certain protein, for which Template 5, *Find proteins which regulate*

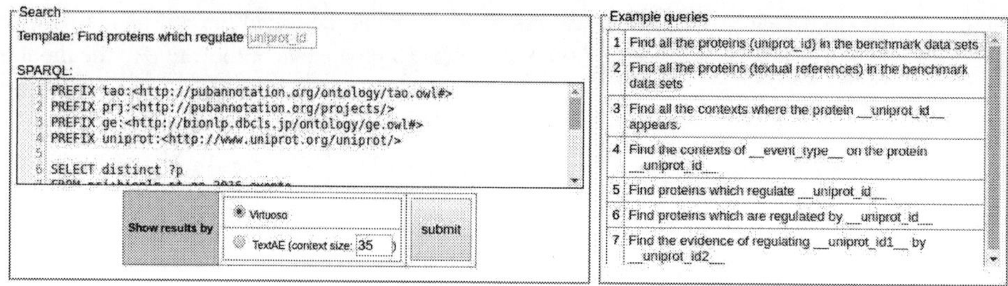

Figure 5: SPARQL interface to the IE-induced KB

```
PREFIX tao:<http://pubannotation.org/ontology/tao.owl#>
PREFIX prj:<http://pubannotation.org/projects/>
PREFIX ge:<http://bionlp.dbcls.jp/ontology/ge.owl#>
PREFIX uniprot:<http://www.uniprot.org/uniprot/>

SELECT DISTINCT ?p
FROM prj:bionlp-st-ge-2016-events
FROM prj:bionlp-st-ge-2016-uniprot
WHERE {
  graph prj:bionlp-st-ge-2016-uniprot {
    ?o1 tao:denoted_by ?s1 .
    ?o1 a uniprot:__uniprot_id__ .
    ?o2 tao:denoted_by ?s2 .
    ?o2 a ?p .
  }

  ?o1_1 tao:denoted_by ?s1 .
  ?o2_1 tao:denoted_by ?s2 .
  ?o1_1 ^ge:partOf? / ge:themeOf+ ?e .
  ?o2_1 ^ge:partOf? / ge:causeOf+ ?e .

  FILTER (?p != tao:Context_entity)
}
```

Figure 6: A SPARQL template of title *Find proteins which regulate __uniprot_id__*

__uniprot_id__, can be used. With *Q04206* (p65) set to the placeholder, we find the following:

- In the ref. data, 21 proteins are found to regulate p65

- In the test data, 2 are found to regulate p65

- Among the 2 proteins found in the test data, one (P01375; TNFα) also in the reference data, whereas the other (P01584; IL1β) only in the test data.

Assuming that the reference data represents a KB at a point, and that the test data represents new feed to the KB, the piece of information that IL1B regulates p65 may represent a new piece of knowledge. On the other hand, the piece of information that TNFα regulates p65 itself may not represent a new knowledge. However, it may supply additional contexts to the known piece of knowledge, from which more detailed information, e.g. experimental condition, may be accessed.

Using Template 7, *Find the evidence for __uniprot_id1__ to regulate __uniprot_id2__*, with P01375 set to the first placeholder, and Q04206 to the second, we can access individual contexts of TNFα to regulate p65. Figure 7 shows one example, which suggests that more detailed knowledge about the regulation may be extracted by further digging the context, e.g., TNFα regulates phosphorylation of p65, and the specific sites of the phosphorylation are Ser529 and Ser536,

The series of analyses demonstrates that how IE results may contribute to populate the KB, and how the IE-driven KB can be explored using the template system.

3.3 Analyses on submissions

Due to the heavy burden of re-implementing the whole task, the GE4 task began as an experimental task. Many problems were encountered during the release of benchmark data sets and the evaluation system, which caused serious delay of the schedule. Thanks to voluntary comments and bug reports from some participants, most of the problems could be addressed, and, in the end, two systems were able to get through to the submission of results. However, as almost no time was given for the participants to adapt their systems to the task, submissions were made using the raw output from the systems, which caused the evaluation scores to be meaninglessly low. Thus, instead of reporting automatic evaluation results, we take the opportunity to discuss observations at the results.

Class	Uniprot ID	Name (Gene)	Reference	Test
I	P19838	Nuclear factor NF-kappa-B p105 subunit (NFKB1)	24	37
	Q00653	Nuclear factor NF-kappa-B p100 subunit (NFKB2)	8	12
II	Q04206	Transcription factor p65 (RELA)	295	98
	Q04864	Proto-oncogene c-Rel (REL)	16	6
	Q01201	Transcription factor RelB (RELB)	6	3

Table 2: Statistics of NFκb proteins in benchmark data sets

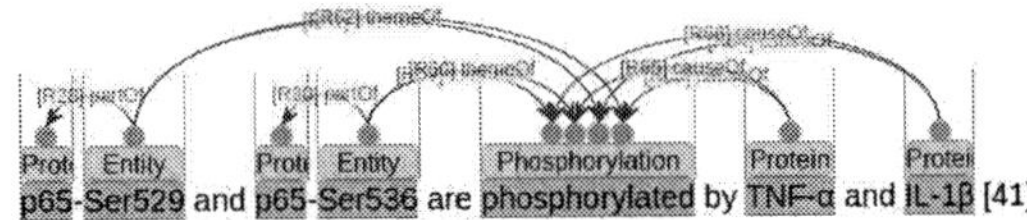

Figure 7: An annotation excerpt from PMC:3312845

One submission was made using the PKDE4J system (Song et al., 2015). An observation on the output revealed that a major discrepancy between the representation of GE4 and the system comes from the fact that while GE4 is an event extraction task PKDE4J is a relation extraction system. In other words, while GE4 requires events to be materialized in the representation, PKDE4J represents them as relations. An example shown in Figure 8 explains the difference. Note that the GE task materializes the events *Negative_regulation* and *Gene_expression* captured by the trigger words *inhibition* and *production*, respectively. While PKDE4J does not materialize them, however, it correctly extracts the relation that *IL-10* down-regulates *interferon gamma*. It also correctly extracts the relation that *IL-10* down-regulates *suppressor of cytokine signaling 1*. Although PKDE4J does not recognize the *Negative_regulation* captured by *Resistance*, it seems right considering that PKDE4J is a relation extraction system which requires two arguments for each relation. The observation suggests that characteristics of individual systems need to be carefully considered to better understand and utilize them.

Furthermore, an attempt was made to use TEES, an open source event extraction system, which won previous iterations of the GE task (Björne and Salakoski, 2013). The goal was to observe TEES' out-of-the-box performance in the GE4 task. With TEES, a different way of entering the task, namely submission of the URL of a RESTful web service, was tested. PubAnnotation offers a function to communicate with a web service to obtain annotations from it. Thus, by submitting the URL of an annotation system which implements a REST-ful API, annotations can be pulled into PubAnnotation. In order to make use of this feature, a small script was written that runs TEES as a RESTful web service, and annotations obtained directly through PubAnnotation. Conversion from the Interaction XML, TEES' native format, to the PubAnnotation JSON format was only minimally implemented, to test the submission. To make use of the performance of TEES, the conversion needs to be implemented more thoroughly, which is left as a future work.

4 Conclusions

The GE4 task is organized as an experimental task, toward generalization of the shared task resources and seamless connection of IE task results to KB population. As the result, a new shared task system is implemented using PubAnnotation as the platform. Note that PubAnnotation itself is an open source project. By being embraced by the open platform, the shared task system is expected to become more sustainable, and accessible. As the newly implemented system is fairly generic, organizing a new shared task is easy, which we hope to promote organization of more shared tasks by interested parties, particularly by domain experts. The GE4 shared task will be running continuously inviting open participation from the community.

Acknowledgments

The work is supported by Life Science Database Integration Project funded by National Bioscience Database Center (NBDC) of Japan Science and Technology Agency (JST).

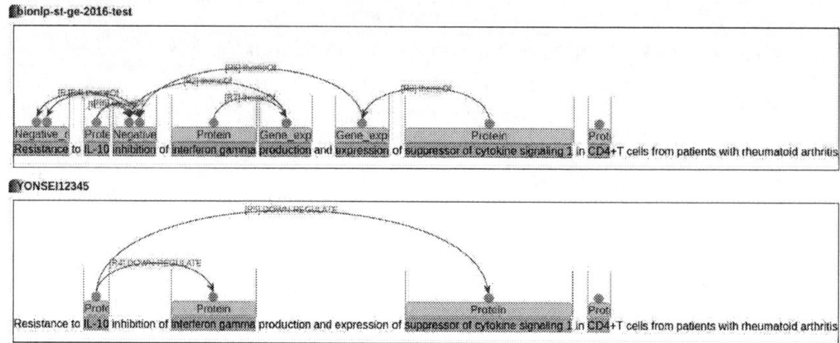

Figure 8: Example of the output of PKDE4J system

References

Jari Björne and Tapio Salakoski. 2013. Tees 2.1: Automated annotation scheme learning in the bionlp 2013 shared task. In *Proceedings of the BioNLP Shared Task 2013 Workshop*, pages 16–25, Sofia, Bulgaria, August.

J Gregory Caporaso, J Lynn Fink, Philip E Bourne, K Bretonnel Cohen, and Lawrence Hunter. 2008. Intrinsic evaluation of text mining tools may not predict performance on realistic tasks. *Pacific Symposium on Biocomputing*, pages 640–6513.

Hong-Woo Chun, Yoshimasa Tsuruoka, Jin-Dong Kim, Rie Shiba, Naoki Nagata, Teruyoshi Hishiki, and Jun'ichi Tsujii. 2006. Extraction of gene-disease relations from medline using domain dictionaries and machine learning. In *Proceedings of the Pacific Symposium on Biocomputing (PSB)*, pages 4–15, Maui, Hawaii, USA, January.

Paolo Ciccarese, Marco Ocana, Leyla Garcia Castro, Sudeshna Das, and Tim Clark. 2011. An open annotation ontology for science on web 3.0. *Journal of Biomedical Semantics*, 2(Suppl 2):S4.

Matthew Honnibal, Yoav Goldberg, and Mark Johnson. 2013. A non-monotonic arc-eager transition system for dependency parsing. In *CoNLL*, pages 163–172.

Jin-Dong Kim and Yue Wang. 2012. Pubannotation: A persistent and sharable corpus and annotation repository. In *Proceedings of the 2012 Workshop on Biomedical Natural Language Processing*, BioNLP '12, pages 202–205, Stroudsburg, PA, USA.

Jin-Dong Kim, Tomoko Ohta, Sampo Pyysalo, Yoshinobu Kano, and Jun'ichi Tsujii. 2009. Overview of BioNLP'09 Shared Task on Event Extraction. In *Proceedings of Natural Language Processing in Biomedicine (BioNLP) Workshop*, pages 1–9.

Jin-Dong Kim, Yue Wang, Toshihisa Takagi, and Akinori Yonezawa. 2011. Overview of the Genia Event task in BioNLP Shared Task 2011. In *Proceedings of the BioNLP 2011 Workshop Companion Volume for Shared Task*, Portland, Oregon, June.

Jin-Dong Kim, Yue Wang, and Yamamoto Yasunori. 2013. The Genia Event Extraction Shared Task, 2013 Edition - Overview. In *Proceedings of the BioNLP Shared Task 2013 Workshop*, pages 8–15, Sofia, Bulgaria, August.

Jin-Dong Kim, Jung-jae Kim, Xu Han, and Dietrich Rebholz-Schuhmann. 2015. Extending the evaluation of genia event task toward knowledge base construction and comparison to gene regulation ontology task. *BMC Bioinformatics*, 16(Suppl 10):S3.

Martin Krallinger, Florian Leitner, and Alfonso Valencia. 2007. Assessment of the Second BioCreative PPI task: Automatic Extraction of Protein-Protein Interactions. In L. Hirschman, M. Krallinger, and A. Valencia, editors, *Proceedings of Second BioCreative Challenge Evaluation Workshop*, pages 29–39.

Kevin Livingston, Michael Bada, Lawrence Hunter, and Karin Verspoor. 2013. Representing annotation compositionality and provenance for the semantic web. *Journal of Biomedical Semantics*, 4(1):38.

Z. Lu, D. Szafron, R. Greiner, P. Lu, D.S. Wishart, B. Poulin, J. Anvik, C. Macdonell, and R. Eisner. 2004. Predicting subcellular localization of proteins using machine-learned classifiers. *Bioinformatics*, 20(4):547–556.

Makoto Miwa, Rune Sætre, Jin-Dong Kim, and Jun'ichi Tsujii. 2010. Event extraction with complex event classification using rich features. *Journal of Bioinformatics and Computational Biology (JBCB)*, 8(1):131–146, February.

Yusuke Miyao and Jun'ichi Tsujii. 2008. Feature forest models for probabilistic hpsg parsing. *Comput. Linguist.*, 34(1):35–80, March.

Christopher J Mungall, Carlo Torniai, Georgios V Gkoutos, Suzanna E Lewis, and Melissa A Haendel. 2012. Uberon, an integrative multi-species anatomy ontology. *Genome Biology*, 13(1):R5.

Min Song, W. C. Kim, D. Lee, G. E. Heo, and K. Y. Kang. 2015. Pkde4j: Entity and relation extraction for public knowledge discovery. *Journal of biomedical informatics*, 57:320–332.

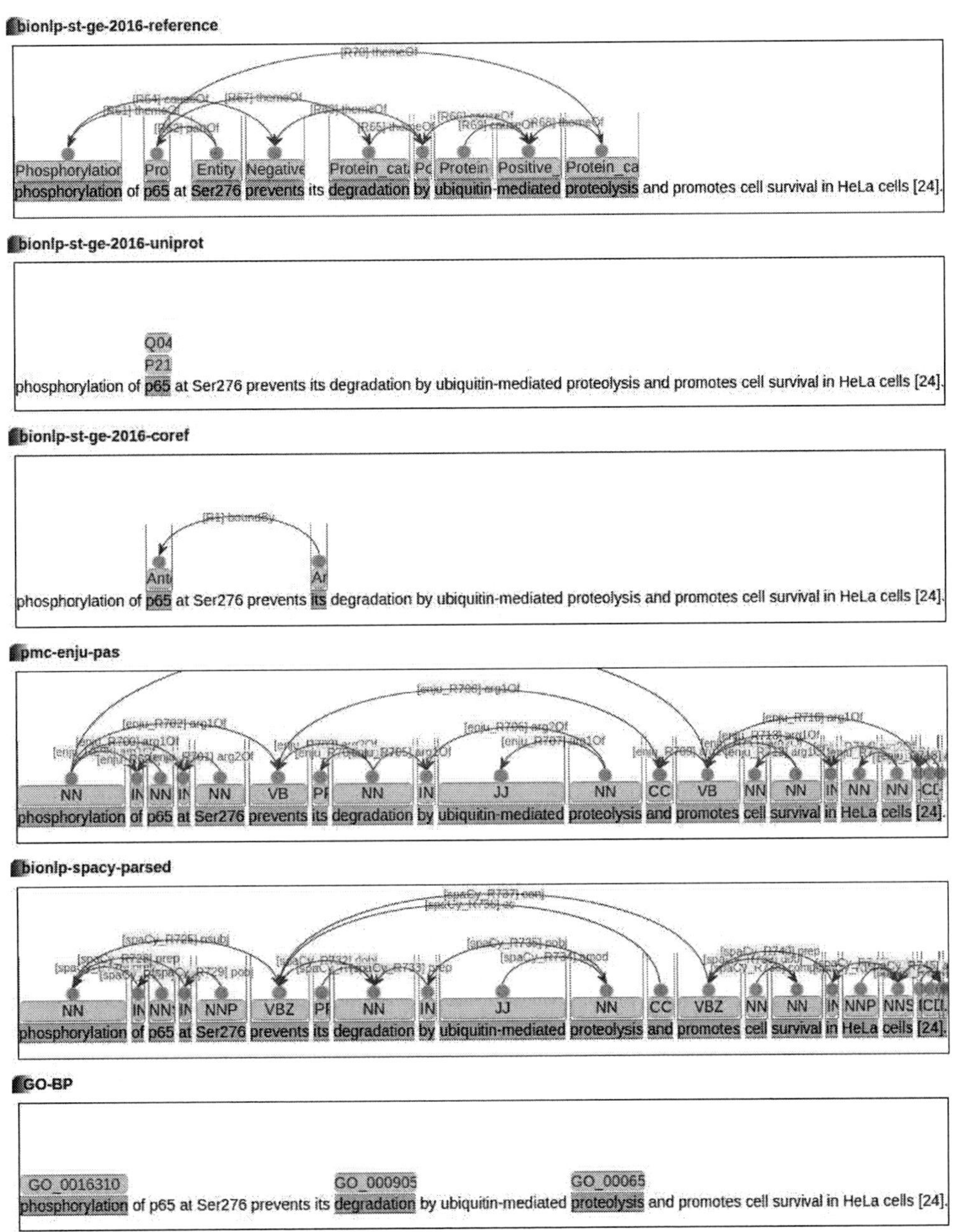

Figure 9: Excerpts of annotation data sets for the GE4 task

LitWay, Discriminative Extraction for Different Bio-Events

Chen Li
Xi'an Jiaotong University, China
Massachusetts Institute of
Technology, United States
`cli@xjtu.edu.cn`

Zhiqiang Rao
Xidian University, China
`zhiqiangrao`
`@foxmail.com`

Xiangrong Zhang
Xidian University, China
Massachusetts Institute of
Technology, United States
`xrzhang`
`@mail.xidian.edu.cn`

Abstract

Even a simple biological phenomenon may introduce a complex network of molecular interactions. Scientific literature is one of the trustful resources delivering knowledge of these networks. We propose LitWay, a system for extracting semantic relations from texts. LitWay utilizes a hybrid method that combines both a rule-based method and a machine learning-based method. It is tested on the SeeDev task of BioNLP-ST 2016, achieves the state-of-the-art performance with the F-score of 43.2%, ranking first of all participating teams. To further reveal the linguistic characteristics of each event, we test the system solely with syntactic rules or machine learning, and different combinations of two methods. We find that it is difficult for one method to achieve good performance for all semantic relation types due to the complication of bio-events in the literatures.

1 Introduction

Bio-events are founding blocks of bio-networks depicting profound biological phenomena. Automatically extracting bio-events may assist researchers while facing the challenge of growing amount of biomedical information in textual form. A bio-event carries more semantic information biochemical reactions between entities, therefore, is more informative for studying associations between bio-concepts, e.g. gene and phenotype (Li et al., 2013).

A number of methods have been proposed to process the automated extraction of biomedical events including rule-based (Cohen et al., 2009; Kilicoglu and Bergler, 2011; Bui and Sloot, 2011)

and machine learning-based (Miwa et al., 2012; Hakala et al., 2013; Munkhdalai et al., 2015) methods. Bui et al. (2013) presented a rule-based method for bio-event extraction by using a dictionary and patterns generated automatically from annotated events. TEES (Björne and Salakoski, 2013) is a SVM based text mining system for the extraction of events and relations from natural language texts, it obtains good performance on a few tasks in BioNLP-ST 2013 (Nédellec et al., 2013). As a major type of biomedical events, a series of methods concentrate on protein-protein interactions (PPI) (Miyao et al., 2009; Papanikolaou et al., 2015). Kernel-based methods are widely used for relation extraction task and obtain good results by leveraging lexical and syntactic information (Airola et al., 2008; Miwa et al., 2009; Li et al., 2015b). Peng et al. (2015) proposed Extended Dependency Graph (EDG) and evaluated it with two kernels on some PPI datasets, obtained good improvements on F-value.

We previously use a set of basic features including word embedding on a classifier for the BioNLP 2013 Genia (Kim et al., 2013) dataset, the result is comparable to the state-of-the-art solution (Li et al., 2015a). The system is built with flexibility in mind. It is designed to tackle more types of bio-events. In this paper, we introduce LitWay, which is based on the previous infrastructure and uses a machine learning based method in combination with syntactic rules. The system is tested on a completely different task, the SeeDev of BioNLP-ST 2016. It achieves the best result among all participants with an F-score of 43.2% (recall and precision are 44.8% and 41.7% respectively).

2 SeeDev Task

As a popular task in unstructured data mining of biomedical interests, BioNLP has successfully

Proceedings of the 4th BioNLP Shared Task Workshop, pages 32–41,
Berlin, Germany, August 13, 2016. ©2016 Association for Computational Linguistics

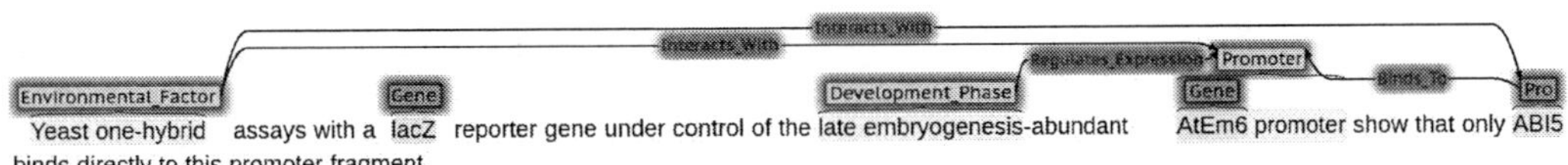

Figure 1: Event relation examples. This sentence includes 4 events and 6 entities. For example, a Environmental_Factor *Yeast one-hybrid* and a Protein *ABI5* form a event Interacts_With. An entity could participate in several events at the same time or none, such as *AtEm6 promoter* and *lacZ*. Noticeably entity span overlap, like Gene *AtEm6* and Promoter *AtEm6 promoter*.

held a series of biomedical event extraction tasks. GE (Genia Event Extraction) is a classic task initiated since the beginning of BioNLP (Kim et al., 2009), it attracts attention and leads to abundant works (Kim et al., 2011; Kim et al., 2013). Be similar to GE and others of BioNLP, SeeDev (Chaix et al., 2016) is a new task proposed in BioNLP-ST 2016, it dedicates to event extraction of genetic and molecular mechanisms involved in plant seed development. It is based on the knowledge model Gene Regulation Network for Arabidopsis (GRNA)[1]. GRNA model defines 16 different types of entities, and 22 event types that may be combined in complex events. Table 1 shows these entities. Event types are presented in following. Figure 1 gives some examples of event relations[2].

Entity type	
Gene	Protein_Domain
Gene_Family	Hormone
Box	Regulatory_Network
Promoter	Pathway
RNA	Genotype
Protein	Tissue
Protein_Family	Development_Phase
Protein_Complex	Environmental_Factor

Table 1: 16 different entity types.

3 Proposed Method

LitWays pipeline adopts a hybrid method that uses a classifier or rule-based method for different event types. Figure 2 shows the infrastructure of it. The pipeline consists of 5 steps: pre-processing, entity pair selection, feature extraction, classifier prediction and rule-based filters.

In BioNLP-ST 2013, the top three event extraction systems F-scores differ less than 0.3%

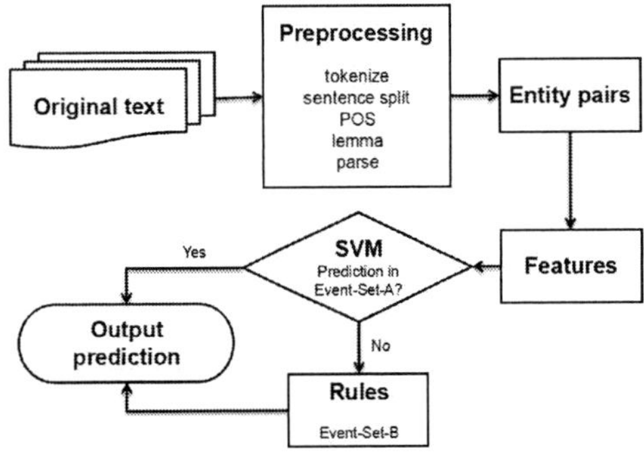

Figure 2: Infrastructure of LitWay.

in number (Nédellec et al., 2013; Björne and Salakoski, 2013). Differences of quantitative and syntactic morphology of proteins and chemical entities in the scientific literature might demand different strategies of network extraction to achieve a better performance. In this paper, we utilize a flexible hybrid system to investigate a way to discriminatively treat event types. We first pre-experiment on the development data and divide all event types into two sets: Event-Set-A and Event-Set-B. The events showing better performance on SVM are classified into Event-Set-A, the others showing better results on a rule-based method are classified into Event-Set-B. Event-Set-A composes of minority events, Event-Set-B composes of majority events except two types: Composes_Primary_Structure and Composes_Protein_Complex[3]. Two sets are showed in Table 2. It is easier to create precise and useful rules for majority events since there are enough instances for analyzing. Compared with using SVM for all events, better results are obtained from the experiment by using the rules.

---

[1]See details at http://2016.bionlp-st.org/tasks/seedev.

[2]From training dataset: SeeDev-binary-11489176-1.

[3]After the analysis with experiment results, while moving Composes_Primary_Structure and Composes_Protein_Complex into Event-Set-A, a slightly better F-score on all events could be obtained.

Event-Set-A	Number
Is_Linked_To	44
Regulates_Accumulation	36
Transcribes_Or_Translates_To	25
Is_Involved_In_Process	23
Occurs_In_Genotype	18
Regulates_Molecule_Activity	16
Exists_At_Stage	15
Regulates_Tissue_Development	9
Occurs_During	8
Event-Set-B	**Number**
Regulates_Process	436
Regulates_Expression	201
Exists_In_Genotype	169
Is_Localized_In	107
Regulates_Development_Phase	106
Is_Member_Of_Family	89
Has_Sequence_Identical_To	62
Interacts_With	62
Is_Functionally_Equivalent_To	60
Binds_To	60
Is_Protein_Domain_Of	46
Composes_Primary_Structure	20
Composes_Protein_Complex	16

Table 2: Event-Set-A and Event-Set-B. This partition was used during competition.

After pre-processing the raw text data, candidate entity pairs are constructed within each sentence, and tested by a multiclass classifier. If the classifier predicts that a candidate pair is a event belonging to Event-Set-A, the predication stays. Otherwise, a series of rules are used for deciding a type in Event-Set-B.

3.1 Pre-processing

The pre-processing include tokenization, sentence splitter, part-of-speech (POS) tagging, lemmatization and syntactic parsing. Stanford CoreNLP tool (Manning et al., 2014) is adopted for the operations.

3.2 Entity Pair Selection

The system aims to resolve semantic relation extraction as expected by the SeeDev task. In the task, each event has two arguments. We construct two entities as a candidate pair each time and predict their relation type. Table 3 presents sentence distance statistics of events on the training set, nearly 96.5% of events span within one sentence.

Since most events occur within a sentence, we only choose entity pairs in the same sentence.

Except three event types Is_Linked_To, Has_Sequence_Identical_To, and Is_Functionally_Equivalent_To, in which two arguments could be reversed, for the others they are ordered. Therefore an entity pair (Entity1, Entity2) is different from the reversed pair (Entity2, Entity1). They should be treated as two instances.

Sentence distance	Number
0	1571
1	52
2	5

Table 3: Sentence distance statistics of events on the training set.

3.3 Feature Extraction

The features are extracted and summarized in Table 4, which shows two types of features, entity features and entity pair features.

Entity feature	Entity pair feature
Entity type	Tree path
Words	Tree path length
Lemmas	Token distance
POSs	Entity distance
Unigram word	Middle lemmas
Unigram lemma	
Unigram POS	
Tree node depth	
Average word embedding	

Table 4: Features used in classifier. Entity features are extracted from two entities, separately. Entity pair features are extracted from a pair of entities.

Word, lemma, Part-Of-Speech (POS) are features directly represent an entity's lexical and grammatical characteristics. Adjacent words' features are used to represent the entity's contextual characteristics. Therefore, basic features include word, lemma, POS of entities, as well as the same information of the unigram words.

Generally speaking, if two entities are closer, they are more likely to be relative (Tikk et al., 2013). Token distance and entity distance are used here. Token distance is the number of tokens between two entities. Entity distance is the number of entities in the middle of two entities.

Syntactic parsing tree features are important for semantic relation (Punyakanok et al., 2008; D'Souza and Ng, 2012). Tree node depth, tree path, tree path length are used in our experiment. They are obtained from the syntactic parsing tree, generated during the pre-processing. Tree node depth is the distance between the corresponding tree node of an entity and the root node of the sentence. Tree path is the path between two entities. Tree path length is the number of middle nodes between two entities in their tree path.

Word embedding has demonstrated the ability of well representing linguistic and semantic information of a text unit (Mikolov et al., 2013; Tang et al., 2014), e.g. POS and N-gram. We continue using it as a feature in our system. Specifically, training, development and test datasets of SeeDev are used to obtain word embedding by using word2vec tool (Mikolov et al., 2013) after sentencization, tokenization and lemmatization on the original text. Since the word number of an entity is uncertain, we use the average value of all the word embeddings of an entity (Chen et al., 2015; Wang et al., 2015), i.e. average word embedding. Middle lemmas include all of the lemmas between two entities, they are treated as a bag-of-word (BOW) feature, some keyword information may be obtained from it.

3.4 SVM Classifier Prediction

Support Vector Machine (SVM) (Cortes and Vapnik, 1995) and the C++ embodiment, LibSVM (Chang and Lin, 2011), is employed for the classification in LitWay. Positive event instances are retrieved from gold annotations. Negative instances are created by all of no-relation entity pairs within each sentence.

Among predication, if the predicted result of an entity pair belongs to Event-Set-A, it is taken as the label. Otherwise, rule-based filters are applied.

3.5 Rule-based Filters

In Event-Set-B, different event types have different rules. We summarize all rules in Table 5. We consider the event types of Event-Set-B one by one, according to their quantities on the training set, as showed in Table 2. Once all rules of an event type are satisfied, the entity pair label could be determined, and the matching of the rest event types could be stopped.

There are 6 types of rules: Event arguments match, Entity structure rules, Sentence structure rules, Token distance restriction, Keywords match and Training set match. The details about these rules are shown as following:

(1) **Event arguments match**: According to the task description, the arguments of the event are strongly typed, which means that all types of entities are not possible as event arguments. What is more, according to the statistics of arguments of different events on the training set, we only retain those arguments that occur most times for each special event type. This could efficiently reduce false instances.

(2) **Entity structure rules**: Many entities have complicated structure, an entity could span over another entity. This results in that some entity structures are less likely to be event arguments. Such as, an entity with smaller span is not an argument, as it is often the modifier of the larger one. Meanwhile, some event types have several fixed special entity argument structures. We summarize 3 particular rules from the training set:

- (2a) Entity is not covered: An entity is not covered by a larger one.

- (2b) Entity does not cover: An entity does not contain smaller entities or overlap with others.

- (2c) Special entity structure: Some special entity structure rules are summarized from the dataset. Presumably an entity pair (Entity1, Entity2) is within a sentence, Entity is another entity in the same sentence, the special entity structures could be:

    - (2c1) *Entity1 (Entity2)*: Entity pair should have such fixed special structure, Entity2 follows Entity1 and is in brackets.
    - (2c2) *Entity1 (Entity)*: If Entity1 is followed by Entity and Entity is in brackets, Entity1 is ignored.
    - (2c3) *Entity (Entity1)*: If Entity1 follows Entity and Entity1 is in brackets, Entity1 is ignored.
    - (2c4) *Entity2 Entity1 (Entity)*: If Entity follows Entity1 and Entity is in brackets, while Entity1 also follows Entity2, then Entity1 is kept.
    - (2c5) *Entity (Entity2)*: If Entity2 follows Entity and Entity2 is in brackets, Entity2 is ignored.

(3) **Sentence structure rules**: If two entities form an event relation, the sentence structure presents some syntactic characteristics. We summarize 3 sentence structure rules:

- (3a) No subordinate clause: Subordinate clause is a complex sentence structure. If there is event relation between a pair of entities, the syntactic tree path structure between them is often simple and direct.

- (3b) Active or passive structure match: For an event argument pair (Entity1, Entity2), it should have such relation structure: *Entity1-influences-Entity2*. While an entity pair has two orders in a sentence: Entity1 is on the left of Entity2, or right of Entity2. Different orders should match different sentence structure rules. If Entity1 is on the left of Entity2, their tree path is an active structure. Otherwise it is a passive structure.

- (3c) Special entity pair order: Some events usually have fixed order between their two arguments, Entity1 is always on the left (or right) of Entity2.

(4) **Token distance restriction**: Closer entities are more likely to be relative. The rule restricts the number of middle tokens between entities. It ignores distant entity pairs.

(5) **Keywords match**: Some events are accompanied by keywords, we record these keywords of several different events, showed in following detailed rules. They are useful for event identification.

(6) **Training set match**: For some event types, we compile the entity pairs from the training set into a dictionary, since they are biologically more likely to interact.

For an entity pair (Entity1, Entity2), we apply the rules on Event-Set-B. Following labels (1) to (6) correspond to 6 type rules introduced above, *None* means nonuse of this rule:

Regulates_Process
(1) Entity1∈{Genotype, Tissue, Gene, Protein, Development_Phase},
Entity2∈{Regulatory_Network, Pathway}
(2) Entity1 is not covered
(3) No subordinate clause, Active or passive structure match
(4) None
(5) None

(1) Event arguments match
(2) Entity structure rules
• (2a) Entity is uncovered
• (2b) Entity does not cover
• (2c) Special entity structure
(3) Sentence structure rules
• (3a) No subordinate clause
• (3b) Active or passive structure match
• (3c) Special entity pair order
(4) Token distance restriction
(5) Keywords match
(6) Training set match

Table 5: Summary of all rules.

(6) None
Regulates_Expression
(1) Entity1∈{Tissue, Genotype, Protein, Development_Phase}, Entity2∈{Gene}
(2) Entity1 is not covered, Entity2 is not covered
(3) No subordinate clause
(4) None
(5) Keywords∈{function, target, repress, bind, regulat-, exclude, activate, require, expression, induce, detect, express, define, act, during, plicate, observe, affect, defect, transcription, cease, associate, restrict, modulate}
(6) None

Exists_In_Genotype
(1) Entity1∈{Gene, Gene_Family, RNA, Protein, Protein_Family, Protein_Domain}
Entity2∈{Genotype}
(2) Entity1 is not covered, Entity2 does not cover
(2c2) *Entity1 (Entity)*: ignore Entity1
(2c4) *Entity2 Entity1 (Entity)*: keep Entity1
(2c5) *Entity (Entity2)*: ignore Entity2
(3) No subordinate clause
(4) None
(5) None
(6) None

Is_Localized_In
(1) Entity1∈{RNA, Protein, Protein_Family, Protein_Complex, Protein_Domain, Hormone}
Entity2∈{Tissue}
(2) Entity1 is not covered, Entity2 is not covered
(3) No subordinate clause
(4) None
(5) None
(6) None

Regulates_Development_Phase
(1) Entity1∈{Gene, Protein, Genotype,

Gene_Family}, Entity2∈{Development_Phase}
(2) Entity1 is not covered, Entity2 is not covered
(2c2) *Entity1 (Entity)*: ignore Entity1
(3) Entity1 is on the left of Entity2
(4) None
(5) None
(6) None

Is_Member_Of_Family
(1) (Entity1, Entity2)∈{(Protein, Protein_Family), (Gene, Gene_Family)}
(2) Entity1 is not covered, Entity2 is not covered
(2c2) *Entity1 (Entity)*: ignore Entity1
(3) No subordinate clause
(4) Token distance ≤ 10
(5) None
(6) None

Has_Sequence_Identical_To
(1) Entity1 and Entity2 have same entity type
(2) (2c1) *Entity1 (Entity2)*: Entity pair has this structure
(3) None
(4) Token distance = 2
(5) None
(6) Training set match

Interacts_With
(1) Entity1∈{Protein, Environmental_Factor},
Entity2∈{Box, Promoter, Protein, Protein_Family, Protein_Complex, Protein_Domain}
(2) Entity1 is not covered, Entity2 is not covered
(3) No subordinate clause
(4) None
(5) Keywords∈{interacts, interacted, interacting, associate, associated, associates, associating}
(6) None

Is_Functionally_Equivalent_To
(1) Entity1 and Entity2 have same entity type
(2) (2c1) *Entity1 (Entity2)*: Entity pair has this structure
(3) None
(4) Token distance = 2
(5) None
(6) Training set match

Binds_To
(1) Entity1∈{Protein, Protein_Family, Protein_Domain}, Entity2∈{Box, Promoter, Protein, Protein_Family, Protein_Complex}
(2) Entity1 is not covered, Entity2 is not covered
(3) No subordinate clause
(4) None
(5) Keywords∈{bind, binds, interact, physical, direct}

(6) None

Is_Protein_Domain_Of
(1) Entity1∈{Protein_Domain},
Entity2∈{Protein, Protein_Family}
(2) Entity1 is not covered, Entity2 is not covered
(2c2) *Entity1 (Entity)*: ignore Entity1
(3) None
(4) None
(5) None
(6) None

Composes_Primary_Structure
(1) Entity1∈{Box}, Entity2∈{Gene, Box, Promoter}
(2) None
(3) No subordinate clause
(4) None
(5) None
(6) None

Composes_Protein_Complex
(1) Entity1∈{Protein},
Entity2∈{Protein_Complex}
(2) Entity1 is not covered, Entity2 is not covered
(3) No subordinate clause, Entity1 is on the left of Entity2
(4) None
(5) None
(6) None

4   Results

To investigate the impact of different strategies and their comparison with the hybrid method, we test the system solely with machine learning, syntactic rules, or different combinations of them.

We compared the proposed hybrid method with the classifier-only based method on the development dataset. Table 6 shows the experiment results. All of the features are beneficial for the classifier, by using all of them we get the best SVM based result with 31.5% F1. Tree features make most improvement with 5.7% increase on F1, both recall and precision are increased. Dist features make only 0.2% F1 improvement and WM features make 1.2% F1 improvement. They increase precision with the loss of recall, while Tree features mainly contribute to recall.

Comparing hybrid method with the best SVM result in Table 6, we could see an obvious advantage. The F1 of the hybrid method is over 10% higher than the best SVM result, it greatly improves recall with around 16%, and has 3.4% precision increase. It's interesting because adding

Method	F1	R	P
(1) Word+POS+Lemma	0.244	0.206	0.300
(2) WPL+Dist	0.246	0.192	0.344
(3) WPL+Dist+Tree	0.303	0.267	0.348
(4) WPL+Dist+Tree+WM	0.315	0.264	0.390
(5) Hybrid	0.423	0.423	0.424

Table 6: Comparison between different features for SVM on development dataset. Methods (1) to (4) only use multiclass SVM with different feature selections, (5) is the hybrid method. WPL are word, POS and lemma features. Dist are token distance and entity distance features. Tree are tree node depth, tree path and tree path length features. WM are average word embedding and middle lemmas features.

rules usually increase precision instead of recall.

To verify the effect of rule-based method for different event types, we take the best SVM result as a basis, and then replace each event type with rule-based method in turns. Event-Set-B uses specific rules introduced before. Event-Set-A uses some frequent rules from Event-Set-B since it is difficult to create precise rules for minority class, they include:
(1) Event arguments match;
(2) Entity1 is not covered, Entity2 is not covered;
(3) No subordinate clause, active or passive structure match.

Table 7 presents the results. Except for Composes_Primary_Structure and Composes_Protein_Complex, F1 of Event-Set-B events are increased by using rules instead of SVM. While rules are not helpful for Event-Set-A, it verifies the partition of two sets.

Since Composes_Primary_Structure and Composes_Protein_Complex have better results in SVM, we move them into Event-Set-A and indeed get a little better result in overall events after the competition, it is showed in following.

Table 8 presents the details of SVM method and hybrid method. Almost all the events of Event-Set-B have better results in the hybrid method. This demonstrates the effectiveness of it.

To investigate the rules used in the proposed method, we take several experiments on the development data by different rule combinations. Table 9 presents their results. All of these rules are beneficial to the system more or less. Event arguments match and entity structure rules have important in-

Method	F1
WPL+Dist+Tree+WM	0.315
Regulates_Process	0.334
Regulates_Expression	0.315
Exists_In_Genotype	0.355
Is_Localized_In	0.323
Regulates_Development_Phase	0.325
Is_Member_Of_Family	0.320
Has_Sequence_Identical_To	0.328
Interacts_With	0.323
Is_Functionally_Equivalent_To	0.330
Binds_To	0.316
Is_Protein_Domain_Of	0.327
Composes_Primary_Structure	0.313
Composes_Protein_Complex	0.313
Is_Linked_To	0.248
Regulates_Accumulation	0.289
Transcribes_Or_Translates_To	0.315
Is_Involved_In_Process	0.306
Occurs_In_Genotype	0.314
Regulates_Molecule_Activity	0.277
Exists_At_Stage	0.313
Regulates_Tissue_Development	0.302
Occurs_During	0.310

Table 7: Replace each event type with rule-based method in turns on the basis of SVM. Event-Set-B is in italic.

fluences to the performance, result in around 10% and 8% F1 decrease respectively. It is understandable because almost all kinds of event types in Event-Set-B use these two rules, which makes them important to the system, especially on the precision. Sentence structure rules and keywords match are also useful, around 3% to 3.5% F1 improvement could be obtained by using them. They improve the performance by increasing the precision of the system with the loss of recall. Token distance restriction and training set match have only 0.1% to 0.3% influences on F1 as they are merely used in one or two event types. Token distance restriction could improve the precision while training set match improves the recall.

Table 10 is the official result of the SeeDev task (Chaix et al., 2016). LitWay achieves the best result among all participating teams with 43.2% F1 showing significant advantage. The recall of LitWay is 44.8%, which is comparable to the highest recall 45.8%. Its precision 41.7% is the second highest value, only lower than 53.3%.

Event type	WPL+Dist+Tree+WM			Hybrid		
	F1	Recall	Precision	F1	Recall	Precision
All	0.315	0.264	0.390	0.423	0.423	0.424
Regulates_Process	0.437	0.447	0.428	0.511	0.520	0.503
Regulates_Expression	0.416	0.387	0.448	0.432	0.360	0.541
Exists_In_Genotype	0.224	0.210	0.239	0.540	0.630	0.472
Is_Localized_In	0.433	0.447	0.420	0.518	0.617	0.446
Regulates_Development_Phase	0.182	0.136	0.276	0.333	0.424	0.275
Is_Member_Of_Family	0.342	0.255	0.519	0.479	0.418	0.561
Has_Sequence_Identical_To	0.514	0.429	0.643	0.807	0.735	0.893
Interacts_With	0.080	0.063	0.111	0.301	0.344	0.268
Is_Functionally_Equivalent_To	0.467	0.318	0.875	0.667	0.590	0.767
Binds_To	0.177	0.125	0.300	0.256	0.208	0.333
Is_Protein_Domain_Of	0.061	0.035	0.250	0.455	0.517	0.405
Composes_Primary_Structure	NA	0	NA	0.238	0.667	0.145
Composes_Protein_Complex	NA	NA	NA	NA	NA	0
Is_Linked_To	0.118	0.087	0.182	0.118	0.087	0.182
Regulates_Accumulation	0.271	0.207	0.429	0.271	0.207	0.429
Transcribes_Or_Translates_To	0.174	0.154	0.200	0.174	0.154	0.200
Is_Involved_In_Process	NA	0	0	NA	0	0
Occurs_In_Genotype	NA	0	NA	NA	0	NA
Regulates_Molecule_Activity	NA	NA	NA	NA	NA	NA
Exists_At_Stage	NA	0	0	NA	0	0
Regulates_Tissue_Development	NA	0	NA	NA	0	NA
Occurs_During	NA	0	NA	NA	0	NA

Table 8: Detailed results of SVM classifier and hybrid method on development dataset. *NA* in Recall represents none of this class instance on the development data. *NA* in Precision represents none of this class instance in the predicted results. *0* in Recall or Precision means none of True Positive (TP) instance of this type is obtained in the predicted results. Event-Set-B is in italic.

We present two more additional experiments after the competition by moving Composes_Primary_Structure and Composes_Protein_Complex into Event-Set-A. Table 11 shows the results. The result on development data has 0.8% improvement on F1, while does not show benefit on test data.

We analyse the results on development dataset before and after the movement operation. Before the movement, for Composes_Primary_Structure there are 10 True Positive (TP) instances among 69 predicted instances (gold number is 15), for composes_Protein_Complex there are 0 TP instance among 8 predicted result (gold number is 0). After the operation both of the two predicted numbers are 0, i.e. we do not make any predictions of the two event types. In this case, 10 right events are lost, on the other hand 67 false events are successfully deleted. It brings more benefits than harm.

5   Conclusion

The paper presents a hybrid method system Lit-Way, to resolve the biomedical semantic relations. It achieves the best result in BioNLP-ST 2016 SeeDev task. It is built as a flexible way with the awareness of that different bio-events have different linguistic characteristics and are difficult to be tackled by a single method.

Without much feature engineering nor complex algorithm, LitWay obtains the state-of-the-art performance on the official test data, with the highest F-score 43.2%. A series of experiments of using the methods and their combinations are carried out to investigate the different linguistic characteristics of different event types.

Here we extract relations within one sentence. While a number of events still span across sentences. By incorporating coreference technics in the future, we expect to be able to interconnect

Method	F1	R	P
(1) Hybrid	0.423	0.423	0.424
(2) No arguments match	0.325	0.484	0.244
(3) No entity rules	0.342	0.447	0.277
(4) No sentence rules	0.394	0.501	0.325
(5) No token distance	0.420	0.423	0.417
(6) No keywords	0.388	0.441	0.347
(7) No training match	0.422	0.420	0.424

Table 9: Hybrid experiment results with different rules on development dataset. Methods (2) to (7) have been removed one type rule separately on the basis of (1). Method (2) only follows the event argument match rules given by the SeeDev task (http://2016.bionlp-st.org/tasks/seedev/seedev-data-representation.), does not filter event arguments that never or rarely occur.

Method	F1	Recall	Precision
LitWay	0.432	0.448	0.417
UniMelb	0.364	0.386	0.345
VERSE	0.342	0.458	0.273
–	0.335	0.245	0.533
ULISBOA	0.306	0.256	0.379
LIMSI	0.255	0.318	0.212
DUTIR*	–	–	–

Table 10: Official evaluation results on test data.

events at the same time improve the event extraction performance.

References

Antti Airola, Sampo Pyysalo, Jari Björne, Tapio Pahikkala, Filip Ginter, and Tapio Salakoski. 2008. All-paths graph kernel for protein-protein interaction extraction with evaluation of cross-corpus learning. *BMC bioinformatics*, 9(11):1.

Jari Björne and Tapio Salakoski. 2013. Tees 2.1: Automated annotation scheme learning in the bionlp 2013 shared task. In *Bionlp Shared Task 2013 Workshop*, pages 16–25.

Quoc-Chinh Bui and Peter Sloot. 2011. Extracting biological events from text using simple syntactic patterns. In *Proceedings of the BioNLP Shared Task 2011 Workshop*, pages 143–146. Association for Computational Linguistics.

Quoc-Chinh Bui, David Campos, Erik van Mulligen, and Jan Kors. 2013. A fast rule-based approach for biomedical event extraction. In *Proceedings of the*

Dataset	F1	Recall	Precision
development	0.431	0.410	0.453
test	0.432	0.439	0.426

Table 11: Additional experiment results.

BioNLP Shared Task 2013 Workshop, pages 104–108. Association for Computational Linguistics.

Estelle Chaix, Bertrand Dubreucq, Abdelhak Fatihi, Dialekti Valsamou, Robert Bossy, Mouhamadou Ba, Louise Deléger, Pierre Zweigenbaum, Philippe Bessières, Loïc Lepiniec, and Claire Nédellec. 2016. Overview of the regulatory network of plant seed development (seedev) task at the bionlp shared task 2016. In *Proceedings of the 4th BioNLP Shared Task workshop*, Berlin, Germany, August. Association for Computational Linguistics.

Chih-Chung Chang and Chih-Jen Lin. 2011. Libsvm: a library for support vector machines. *ACM Transactions on Intelligent Systems and Technology (TIST)*, 2(3):27.

Xinxiong Chen, Lei Xu, Zhiyuan Liu, Maosong Sun, and Huanbo Luan. 2015. Joint learning of character and word embeddings. In *Proceedings of IJCAI*, pages 1236–1242.

K Bretonnel Cohen, Karin Verspoor, Helen L Johnson, Chris Roeder, Philip V Ogren, William A Baumgartner Jr, Elizabeth White, Hannah Tipney, and Lawrence Hunter. 2009. High-precision biological event extraction with a concept recognizer. In *Proceedings of the Workshop on Current Trends in Biomedical Natural Language Processing: Shared Task*, pages 50–58. Association for Computational Linguistics.

Corinna Cortes and Vladimir Vapnik. 1995. Support-vector networks. *Machine Learning*, 20(3):273—297.

Jennifer D'Souza and Vincent Ng. 2012. Anaphora resolution in biomedical literature: A hybrid approach. In *Proceedings of the ACM Conference on Bioinformatics, Computational Biology and Biomedicine*, pages 113–122. ACM.

Kai Hakala, Sofie Van Landeghem, Tapio Salakoski, Yves Van de Peer, and Filip Ginter. 2013. Evex in st13: Application of a large-scale text mining resource to event extraction and network construction. In *Proceedings of the BioNLP Shared Task 2013 Workshop*, pages 26–34. Association for Computational Linguistics.

Halil Kilicoglu and Sabine Bergler. 2011. Adapting a general semantic interpretation approach to biological event extraction. In *Proceedings of the BioNLP Shared Task 2011 Workshop*, pages 173–182. Association for Computational Linguistics.

Jin-Dong Kim, Tomoko Ohta, Sampo Pyysalo, Yoshinobu Kano, and Jun'ichi Tsujii. 2009. Overview of bionlp'09 shared task on event extraction. In *Proceedings of the Workshop on Current Trends in Biomedical Natural Language Processing: Shared Task*, pages 1–9. Association for Computational Linguistics.

Jin-Dong Kim, Yue Wang, Toshihisa Takagi, and Akinori Yonezawa. 2011. Overview of genia event task in bionlp shared task 2011. In *Proceedings of the BioNLP Shared Task 2011 Workshop*, pages 7–15. Association for Computational Linguistics.

Jin-Dong Kim, Yue Wang, and Yamamoto Yasunori. 2013. The genia event extraction shared task, 2013 edition-overview. In *Proceedings of the BioNLP Shared Task 2013 Workshop*, pages 8–15. Association for Computational Linguistics.

Chen Li, Maria Liakata, and Dietrich Rebholzschuhmann. 2013. Biological network extraction from scientific literature: State of the art and challenges. *Briefings in Bioinformatics*, 15(5):856–877.

Chen Li, Runqing Song, Maria Liakata, Andreas Vlachos, Stephanie Seneff, and Xiangrong Zhang. 2015a. Using word embedding for bio-event extraction. *ACL-IJCNLP 2015*, page 121.

Lishuang Li, Rui Guo, Zhenchao Jiang, and Degen Huang. 2015b. An approach to improve kernel-based protein–protein interaction extraction by learning from large-scale network data. *Methods*, 83:44–50.

Christopher D Manning, Mihai Surdeanu, John Bauer, Jenny Rose Finkel, Steven Bethard, and David McClosky. 2014. The stanford corenlp natural language processing toolkit. In *ACL (System Demonstrations)*, pages 55–60.

Tomas Mikolov, Ilya Sutskever, Kai Chen, Greg S Corrado, and Jeff Dean. 2013. Distributed representations of words and phrases and their compositionality. In *Advances in neural information processing systems*, pages 3111–3119.

Makoto Miwa, Rune Sætre, Yusuke Miyao, and Junichi Tsujii. 2009. Protein–protein interaction extraction by leveraging multiple kernels and parsers. *International journal of medical informatics*, 78(12):e39–e46.

Makoto Miwa, Paul Thompson, and Sophia Ananiadou. 2012. Boosting automatic event extraction from the literature using domain adaptation and coreference resolution. *Bioinformatics*, 28(13):1759–1765.

Yusuke Miyao, Kenji Sagae, Rune Sætre, Takuya Matsuzaki, and Jun'ichi Tsujii. 2009. Evaluating contributions of natural language parsers to protein–protein interaction extraction. *Bioinformatics*, 25(3):394–400.

Tsendsuren Munkhdalai, Oyun-Erdene Namsrai, and Keun H Ryu. 2015. Self-training in significance space of support vectors for imbalanced biomedical event data. *BMC bioinformatics*, 16(Suppl 7):S6.

Claire Nédellec, Robert Bossy, Jin Dong Kim, Jung Jae Kim, Tomoko Ohta, Sampo Pyysalo, and Pierre Zweigenbaum. 2013. Overview of bionlp shared task 2013. In *Bionlp Shared Task 2013 Workshop*, pages 1–7.

Nikolas Papanikolaou, Georgios A Pavlopoulos, Theodosios Theodosiou, and Ioannis Iliopoulos. 2015. Protein–protein interaction predictions using text mining methods. *Methods*, 74:47–53.

Yifan Peng, Samir Gupta, Cathy H Wu, and K Vijay-Shanker. 2015. An extended dependency graph for relation extraction in biomedical texts. *ACL-IJCNLP 2015*, page 21.

Vasin Punyakanok, Dan Roth, and Wen-tau Yih. 2008. The importance of syntactic parsing and inference in semantic role labeling. *Computational Linguistics*, 34(2):257–287.

Duyu Tang, Furu Wei, Nan Yang, Ming Zhou, Ting Liu, and Bing Qin. 2014. Learning sentiment-specific word embedding for twitter sentiment classification. In *ACL (1)*, pages 1555–1565.

Domonkos Tikk, Illés Solt, Philippe Thomas, and Ulf Leser. 2013. A detailed error analysis of 13 kernel methods for protein–protein interaction extraction. *BMC bioinformatics*, 14(1):1.

Huazheng Wang, Bin Gao, Jiang Bian, Fei Tian, and Tie-Yan Liu. 2015. Solving verbal comprehension questions in iq test by knowledge-powered word embedding. *arXiv preprint arXiv:1505.07909*.

VERSE: Event and Relation Extraction in the BioNLP 2016 Shared Task

Jake Lever and Steven JM Jones
Canada's Michael Smith Genome Sciences Centre
570 W 7th Ave, Vancouver
BC, V5Z 4S6, Canada
{jlever,sjones}@bcgsc.ca

Abstract

We present the Vancouver Event and Relation System for Extraction (VERSE)[1] as a competing system for three subtasks of the BioNLP Shared Task 2016. VERSE performs full event extraction including entity, relation and modification extraction using a feature-based approach. It achieved the highest F1-score in the Bacteria Biotope (BB3) event subtask and the third highest F1-score in the Seed Development (SeeDev) binary subtask.

1 Introduction

Extracting knowledge from biomedical literature is a huge challenge in the natural language parsing field and has many applications including knowledge base construction and question-answering systems. Event extraction systems focus on this problem by identifying specific events and relations discussed in raw text.

Events are described using three key concepts, entities, relations and modifications. Entities are spans of text that describe a specific concept (e.g. a gene). Relations describe a specific association between two (or potentially more) entities. Together entities and relations describe an event or set of events (such as complex gene regulation). Modifications are changes made to events such as speculation.

The BioNLP Shared Tasks have encouraged research into new techniques for a variety of important NLP challenges. Occurring in 2009, 2011 and 2013, the competitions were split into several subtasks (Kim et al., 2009; Kim et al., 2011; Nédellec et al., 2013). These subtasks provided annotated texts (commonly abstracts from PubMed) of entities, relations and events in a particular biomedical

domain. Research groups were then challenged to generate new tools to better predict new relations and events in test data.

The BioNLP 2016 Shared Task contains three separate parts, the Bacteria Biotope subtask (BB3), the Seed Development subtask (SeeDev) and the Genia Event subtask (GE4). The BB3 and SeeDev subtasks have separate parts that specialise in entity recognition and relation extraction. The GE4 subtask focuses on full event extraction of NFkB related gene events.

Previous systems for relation and event extraction have taken two main approaches: rule-based and feature-based. Rule-based methods learn specific patterns that fit different events, for instance, the word "expression" following a gene name generally implies an expression event for that gene. The pattern-based tool BioSem (Bui et al., 2013) in particular performed well in the Genia Event subtask of the BioNLP'13 Shared Task. Feature-based approaches translate the textual content into feature vectors that can be analysed with a traditional classification algorithm. Support vector machines (SVMs) have been very popular with successful relation extraction tools such as TEES (Björne and Salakoski, 2013).

We present the Vancouver Event and Relation System for Extraction (VERSE) for the BB3 event subtask, the SeeDev binary subtask and the Genia Event subtask. Utilising a feature-based approach, VERSE builds on the ideas of the TEES system. It offers control over the exact semantic features to use for classification, allows feature selection to reduce the size of feature vectors and uses a stochastic optimisation strategy with k-fold cross-validation to identify the best parameters. We examine the competitive results for the various subtasks and also analyse VERSE's capability to predict relations across sentence boundaries.

---

[1] http://www.github.com/jakelever/VERSE

Proceedings of the 4th BioNLP Shared Task Workshop, pages 42–49,
Berlin, Germany, August 13, 2016. ©2016 Association for Computational Linguistics

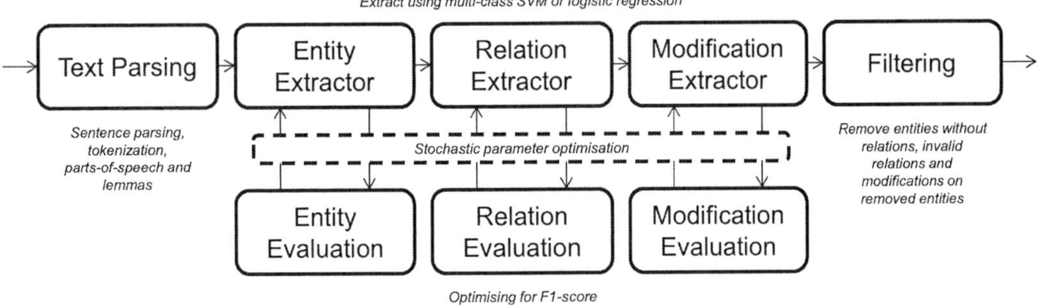

Figure 1: Overview of VERSE pipeline

2 Pipeline

VERSE breaks event extraction into five steps outlined in the pipeline shown in Figure 1. Firstly the input data is passed through a text processing tool that splits and tags text and associates the parsed results with the provided annotations. This parsed data is then passed through three separate classifications steps for entities, relations and modifications. Finally, the results are filtered to make sure that all relations and modifications fit the expected types for the given task.

2.1 Text processing

VERSE can accept input in the standard BioNLP-ST format or the PubAnnotation JSON format (Kim and Wang, 2012). Both formats are standoff, as they contain the text and annotations separately. The annotations describe entities in the text as spans of text and relations and modifications of these entities.

The input files for the shared subtasks are initially processed using the Stanford CoreNLP toolset. The texts are split into sentences and tokenized. Parts-of-speech and lemmas are identified and a dependency parse is generated for each sentence. CoreNLP also returns the exact positions of each token. Using this data, an interval tree is created to identify intersections of text with entities described in the associated annotation. The specific sentence and locations of each associated word are then stored for each entity. Relations and modifications described in the associated annotations are also loaded, retaining information on which entities are involved.

The entities in the BB3 and SeeDev subtasks are generally sets of full words but can be non-contiguous. Entities are stored as a set of associated words rather than a span of words. The GE4 task also contains entities that contain only partial words, for example, "PTEN" is tagged as an entity within "PTEN-deficient". A list of common prefixes and suffixes from the GE4 task is used to separate these words into two words so that the example would become "PTEN deficient". Furthermore, any annotation that divides a word that contains a hyphen or forward slash causes the word to be separate into two separate words.

For easier interoperability, the text parsing code was developed in Jython (Developers, 2008) (a version of Python that can load Java libraries, specifically the Stanford CoreNLP toolset). This Jython implementation is then able to export easily processed Python data structures. Due to incompatibility between Jython and various numerical libraries, a separate Python-only implementation loads the generated data structures for further processing and classification.

2.2 Candidate generation

For all three classifications steps (entities, relations and modifications), the same machine learning framework is used. All possibles candidates are generated for entities, relations or modifications. For relations, this means all pairs of entities are found (within a certain sentence range). For the training step, the candidates are associated with a known class (i.e. the type of relation), or the negative class if the candidate is not annotated in the training set. For testing, the classes are unknown. Candidates can contain one argument (for entity extraction and modification) or two arguments (for relation extraction). These arguments are stored as references to sentences and the indices of the associated words.

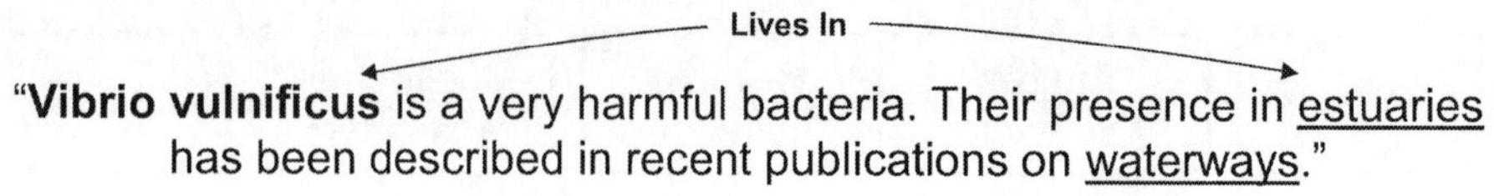

Lives_In	Argument 1	Argument 2
No	vibrio vulnificus	waterways
Yes	vibrio vulnificus	estuaries
No	waterways	vibrio vulnificus
No	waterways	waterways
No	estuaries	vibrio vulnificus
No	estuaries	estuaries

(a) Ignoring argument types

Lives_In	Argument 1 (Bacteria)	Argument 2 (Habitat)
No	vibrio vulnificus	waterways
Yes	vibrio vulnificus	estuaries

(b) Filtering using argument types

Figure 2: Relation candidate generation for the example text which contains a single Lives_In relation (between bacteria and habitat). The bacteria entity is shown in bold and the habitat entities are underlined. Relation example generation creates pairs of entities that will be vectorised for classification. (a) shows all pairs matching without filtering for specific entity types (b) shows filtering for entity types of bacteria and habitat for a potential Lives_In relation

2.2.1 Entity extraction

Entity extraction aims to classify individual or sets of words as a certain type of entity, given a set of training cases. Entities may contain non-contiguous words. The set of all possible combinations of words that could compose an entity is too large for the classification system. Hence VERSE filters for only combinations of words that are identified as entities in the training set.

2.2.2 Relation extraction

VERSE can predict relations between two entities, also known as binary relations. Candidates for each possible relation are generated for every pair of entities that are within a fixed sentence range. Hence when using the default sentence range of 0, only pairs of entities within the same sentence are analysed. VERSE can optionally filter pairs of entities using the expected types for a set of relations as shown in Figure 2.

Each candidate is linked with the locations of the two entities. If the two entities are already annotated to be in a relation, then they are labelled with the corresponding class. Otherwise, the binary relation candidate is annotated with the negative class.

2.2.3 Modification extraction

VERSE supports modification of entities in the form of event modification but currently does not support modification of individual relations. A modification candidate is created for all entities that form the base of an event. These entities are often known as the triggers of the event. In the JSON format, these entities traditionally have IDs that start with "E". If a modification exists in the training set for that entity, the appropriate class is associated with it. Individual binary classifiers are generated for each modification type. This allows an event to be classified with more than one modification.

2.3 Classification

All candidates are vectorized using the same framework, whether for candidates with one or two arguments with minor changes. The full set of features is outlined in Section 3. These vectorized candidates are then used for training a traditional classifier. The vectors may be reduced using feature selection. Most importantly, the parameters used for the feature generation and classifier can easily be varied to find the optimal results. Classification uses the scikit-learn Python package (Pedregosa et al., 2011).

2.3.1 Feature selection

VERSE implements optional feature selection using a chi-squared test on individual parameters against the class variable. The highest ranking features are then filtered based on the percentage of features desired.

2.3.2 Classifier parameters

Classification uses either a support vector machine (SVM) or logistic regression. When using the SVM, the linear kernel is used due to lower time complexity. The multi-class classification uses a one-vs-one approach. The additional parameters of the SVM that are optimised are the penalty parameter C, class weighting approach and whether to use the shrinking heuristic. The class weighting is important as the negative samples greatly outnumber the positive samples for most problems.

2.3.3 Stochastic parameter optimisation

VERSE allows adjustment of the various parameters including the set of features to generate, the classifier to use and the associated classification parameters. The optimisation strategy involves initially seeding 100 random parameter sets. After this initial set, the top 100 previous parameter sets are identified each iteration and one is randomly selected. This parameter set is then tweaked as follows. With a probability of 0.05, an individual parameter is changed. In order to avoid local maxima, an entirely new parameter set is generated with a probability of 0.1. For the subtasks, a 500 node cluster using Intel X5650s was used for optimisation runs.

The optimal parameters are determined for the entity extraction, relation extraction and each possible modification individually. In order to balance precision and recall equally at each stage, the F1-score is used.

2.4 Filtering

Final filtering is used to remove any predictions that do not fit into the task specification. Firstly all relations are checked to see that the types of the arguments are appropriate. Any entities that are not included in relations are removed. Finally, any modifications that do not have appropriate arguments or were associated with removed entities are also trimmed.

Feature Name	Target
unigrams	Entire Sentence
unigrams & parts-of-speech	Entire Sentence
bigrams	Entire Sentence
skipgrams	Entire Sentence
path edges type	Dependency Path
unigrams	Dependency Path
bigrams	Dependency Path
unigrams	Each Entity
unigrams & parts-of-speech	Each Entity
nearby path edge types	Each Entity
lemmas	Each Entity
entity types	Each Entity
unigrams of windows	Each Entity
is relation across sentences	N/A

Table 1: Overview of the various features that VERSE can use for classification

2.5 Evaluation

An evaluation system was created that generates recall, precision, and associated F1-scores for entities, relations and modifications. The system works conservatively and requires exact matches. It should be noted that our internal evaluation system gave similar but not exactly matching results to the online evaluation system for the BB3 and SeeDev subtasks.

K-fold cross-validation is used in association with this evaluation system to assess the success of the system. The entity, relation and modification extractors are trained separately. For the BB3 and SeeDev subtasks, two-fold cross-validation is used, using the provided split of training and development sets as the training sets for the first and second fold respectively. For the GE4 task, five-fold cross-validation is used. The average F1-score of the multiple folds is used as the metric of success.

3 Features

For each generated candidate, a variety of features (controllable through a parameter) is calculated. The features focus on characteristics of the full sentence, dependency path or individual entities. The full-set is shown in Table 1. It should also be noted that a term frequency-inverse document frequency (TFIDF) transform is also an option for all bag-of-words based features.

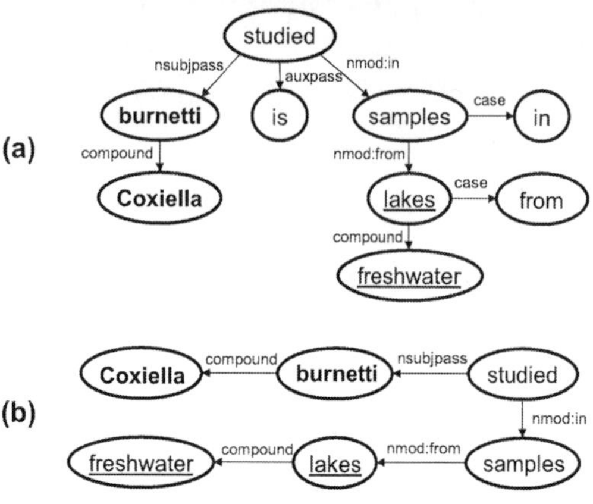

Figure 3: Dependency parsing of the shown sentence provides (a) the dependency graph of the full sentence which is then reduced to (b) the dependency path between the two multi-word terms. This is achieved by finding the subgraph which contains all entity nodes and the minimum number of additional nodes.

3.1 Full sentence features

N-grams features (unigrams and bigrams) use a bag-of-words approach to count the word occurrences across the whole sentence. The words are transformed to lowercase but notably are not filtered for stop words. A version combining the individual words with part-of-speech information is also used. A bag-of-words vector is also generated for lemmas of all words in the sentence. Skip-gram-like features are generated using two words separated by a fixed window of words are also used to generate features. Hence the terms "regulation of EGFR" and "regulation with EGFR" would match the same features of "regulation * EGFR".

3.2 Dependency path features

The dependency path is the shortest path between the two entities in a dependency parse graph and has been shown to be important for relation extraction (Bunescu and Mooney, 2005). Features generated from the set of edges and nodes of the dependency graph include a unigrams and bigrams representation. The specific edge types in the dependency path are also captured with a bag-of-words vector. In order to give specific information about the location of the entity in the dependency path, the types of the edges leaving the entity nodes are recorded separately for each entity.

Interestingly an entity may span multiple nodes in the dependency graph. An example of a dependency path with the multi-word entities "coxiella burnetii" and "freshwater lakes" is shown in Figure 3. In this case, the minimal subgraph that connects all entity nodes in the graph is calculated. This problem was transformed into a minimal spanning tree problem as follows and solved using the NetworkX Python package (Hagberg et al., 2008). The shortest paths through the graph were found for all pairs of entity nodes (nodes associated with the multi-word entities). The path distance between each pair was totalled and used to generate a new graph containing only the entity nodes. The minimal spanning tree was calculated and the associated edges recovered to generate the minimal subgraph. This approach would allow for a dependency path-like approach for relations between more than two entities.

3.3 Entity features

The individual entities are also used to generate specific features. Three different vectorised versions use a unigrams approach, a unigrams approach with parts-of-speech information and lemmas respectively. A one-hot vector approach is used to represent the type of each entity. Unigrams of words around each entity within a certain window size are also generated.

3.4 Multi-sentence and single entity features

VERSE is also capable of generating features for relations between two entities that are in different sentences. In this case, all sentence features are generated for both sentences together and no changes are made to the entity features.

The dependency path features are treated differently. The dependency path for each entity is created as the path from the entity to the root of the dependency graph, generally the main verb of the sentence. This then creates two separate paths, one per sentence and the features are generated in similar ways using these paths. Finally, a simple binary feature is generated for relation candidates that span multiple sentences.

For relation and modifications, candidates contain only a single argument. The dependency path is created in a similar manner to candidates of relations that span across sentences.

Parameter	BB3 event	SeeDev binary
Features	unigrams unigrams POS bigrams of dependency path unigrams of dependency path path edges types entity types entity lemmas entity unigrams POS path edges types near entities	unigrams unigrams POS path edges types path edges types near entities entity types
Feature Selection	No	Top 5%
Use TFIDF	Yes	Yes
Sentence Range	0	0
SVM Kernel	linear	linear
SVM C Parameter	0.3575	1.0 (default)
SVM Class Weights	Auto	5 for positive and 1 for negative
SVM Shrinking	No	No

Table 2: Parameters used for BB3 and SeeDev subtasks

4 Results and discussion

The VERSE tool as described was applied to three subtasks: the BB3 event subtask, the SeeDev binary subtask and the GE4 subtask.

4.1 Datasets

The BB3 event dataset provided by the BioNLP-ST 16 organizers contains a total of 146 documents (with 61, 34 and 51 documents in the training, development and test sets respectively). These documents are annotated with entities of the following types and associated total counts: bacteria (932), habitat (1,861) and geographical (110). Only a single relation type (Lives_In) is annotated which must be between a bacteria and habitat or a bacteria and a geographical entity.

The dataset for the SeeDev binary subtask contains 20 documents with a total of 7,082 annotated entities and 3,575 relations. There are 16 entity types and 22 relation types.

The GE4 dataset focuses on NFkB gene regulation and contains 20 documents. After filtering for duplicates and cleanup, it contains 13,012 annotated entities of 15 types. These entities are in 7,232 relations of 5 different types. It also contains 81 negation and 121 speculation modifications for events. Coreference data is also provided but was not used.

4.2 Cross-validated results

Both BB3 event and SeeDev binary subtasks required only relation extraction. VERSE was trained through cross-validation using the parameter optimising strategy and the optimal parameters are outlined in Table 2. Both tasks were split into training and development sets by the competition organisers. The training set contained roughly twice as many annotations as the development set. We used this existing split for the two-fold cross-validation. A linear kernel SVM was found to perform the best in both tasks. For both subtasks, relation candidates were generated ignoring the argument types as shown in Figure 2.

The classifiers for the two tasks use two very different sizes of feature vectors. The BB3 parameter set has a significant amount of repeated unigrams data, with unigrams for the dependency path and whole sentence with and without parts of speech. This parameter set also does not use feature selection, meaning that the feature vectors are very large (14,862 features). Meanwhile, the SeeDev parameters use feature selection to select the top 5% of features which reduces the feature vector from 7,140 features down to only 357. This size difference is very interesting and warrants further exploration of feature selection for other tasks.

Unfortunately, both classifiers performed best with a sentence range of zero, meaning that only relations within sentences could be detected. Ta-

	Fold 1	Fold 2	Average
Recall	0.552	0.610	0.581
Precision	0.469	0.582	0.526
F1-score	0.507	0.596	0.552

Table 3: Cross-validated results of BB3 event subtask using optimal parameters in Table 2

	Fold 1	Fold 2	Average
Recall	0.363	0.386	0.375
Precision	0.261	0.246	0.254
F1-score	0.303	0.301	0.302

Table 4: Cross-validated results of SeeDev binary subtask using optimal parameters in Table 2

	BB3 event	SeeDev binary
Recall	0.615	0.458
Precision	0.510	0.273
F1-score	0.558	0.342

Table 6: Final reported results for the BB3 event and SeeDev binary subtasks

	Entities	Relations	Mods
Recall	0.71	0.23	0.11
Precision	0.94	0.60	0.38
F1-score	0.81	0.33	0.17

Table 7: Final reported results for GE4 subtask split into entity, relations and modifications results

bles 3 and 4 show the optimal cross-validated results that were found with these parameters. Notably, the F1-scores for the two folds of the SeeDev dataset are very similar, which is surprising given that the datasets are different sizes.

For the GE4 subtask, the cross-validation based optimisation strategy was used to find parameters for the entity, relation and modification extractions independently. Due to the larger dataset, filtering was applied to the argument types of relation candidates as shown in Figure 2. Table 5 outlines the resulting F1-scores from the five-fold cross-validations. As these extractors are trained separately, their performance in the full pipeline would be expected to be worse. This is explained as any errors during entity extraction are passed onto relation and modification extraction.

4.3   Competition results

The official results for the BB3 and SeeDev tasks are shown in Table 6. VERSE performed well in both tasks and was ranked first for the BB3 event subtask and third for the SeeDev binary subtask. The worse performance for the SeeDev dataset may be explained by the added complexity of many additional relation and entity types.

Table 7 shows the final results for the test set

	Entities	Relations	Mods
Recall	0.703	0.695	0.374
Precision	0.897	0.736	0.212
F1-score	0.786	0.715	0.266

Table 5: Averaged cross-validated F1-score results of GE4 event subtask with entities, relations and modifications trained separately

for the Genia Event subtask using the online evaluation tool. As expected, the F1-scores of the relation and modification extraction are lower for the full pipeline compared to the cross-validated results. Nevertheless, the performance is very reasonable given the more challenging dataset.

4.4   Multi-sentence analysis

29% of relations span sentence boundaries in the BB3 event dataset and 4% in the SeeDev dataset. Most relation extraction systems do not attempt to predict these multi-sentence relations. Given the higher proportion in the BB3 set, we use this dataset for further analysis of VERSE's ability to predict relations that span sentence boundaries. It should be noted that some of these relations may be artifacts due to false identification of sentence boundaries by the CoreNLP pipeline.

Using the optimal parameters for the BB3 problem, we analysed prediction results using different values for the sentence range parameter. The performance, shown in Figure 4, is similar for relations within the same sentence using different sentence range parameters. However, as the distance of the relation increases, the classifier predicts larger ratios of false positives to true positives. With sentence range = 3, the overall F1-score for the development set has dropped to 0.326 from 0.438 when sentence range = 1.

The classifier is limited by the small numbers of multi-sentence relations to use as a training set. With a suitable amount of data, it would be worthwhile exploring the use of separate classifiers for relations that are within sentences and those that span sentences as they likely depend on different features.

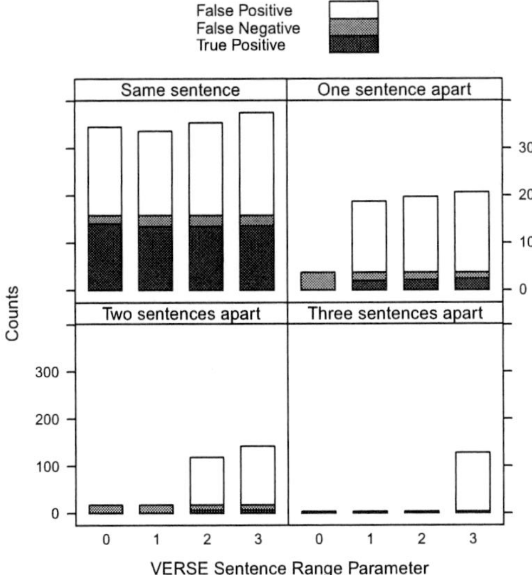

Figure 4: Analysis of performance on binary relations that cross sentence boundaries. The classifier was trained on the BB3 event training set and evaluated using the corresponding development set.

5   Conclusion

We have presented VERSE, a full event extraction system that performed very well in the BioNLP 2016 Shared Task. The VERSE system builds upon the success of previous systems, particularly TEES, in several important ways. It gives full control of the specific semantic features used to build the classifier. In combination with the stochastic optimisation strategy, this control has been shown to be important given the differing parameter sets found to be optimal for the different subtasks. Secondly, VERSE allows for feature selection which is important in reducing the size of the large sparse feature vectors and avoid overfitting. Lastly, VERSE can predict relations that span sentence boundaries, which is certain to be an important avenue of research for future relation extraction tasks. We hope that this tool will become a valuable asset in the biomedical text-mining community.

Acknowledgments

This research was supported by a Vanier Canada Graduate Scholarship. The authors would like to thank ComputeCanada for computational resources used in this project.

References

Jari Björne and Tapio Salakoski. 2013. TEES 2.1: Automated annotation scheme learning in the BioNLP 2013 Shared Task. In *Proceedings of the BioNLP Shared Task 2013 Workshop*, pages 16–25.

Quoc-Chinh Bui, David Campos, Erik van Mulligen, and Jan Kors. 2013. A fast rule-based approach for biomedical event extraction. In *Proceedings of the BioNLP Shared Task 2013 Workshop*, pages 104–108. Association for Computational Linguistics.

Razvan C Bunescu and Raymond J Mooney. 2005. A shortest path dependency kernel for relation extraction. In *Proceedings of the conference on human language technology and empirical methods in natural language processing*, pages 724–731. Association for Computational Linguistics.

Jython Developers. 2008. Jython implementation of the high-level, dynamic, object-oriented language python written in 100% pure Java. Technical report, Technical report (1997-2016), http://www. jython. org/(accessed May 2016).

Aric A. Hagberg, Daniel A. Schult, and Pieter J. Swart. 2008. Exploring network structure, dynamics, and function using NetworkX. In *Proceedings of the 7th Python in Science Conference (SciPy2008)*, pages 11–15, Pasadena, CA USA, August.

Jin-Dong Kim and Yue Wang. 2012. PubAnnotation: a persistent and sharable corpus and annotation repository. In *Proceedings of the 2012 Workshop on Biomedical Natural Language Processing*, pages 202–205. Association for Computational Linguistics.

Jin-Dong Kim, Tomoko Ohta, Sampo Pyysalo, Yoshinobu Kano, and Jun'ichi Tsujii. 2009. Overview of BioNLP'09 shared task on event extraction. In *Proceedings of the Workshop on Current Trends in Biomedical Natural Language Processing: Shared Task*, pages 1–9. Association for Computational Linguistics.

Jin-Dong Kim, Sampo Pyysalo, Tomoko Ohta, Robert Bossy, Ngan Nguyen, and Jun'ichi Tsujii. 2011. Overview of BioNLP shared task 2011. In *Proceedings of the BioNLP Shared Task 2011 Workshop*, pages 1–6. Association for Computational Linguistics.

Claire Nédellec, Robert Bossy, Jin-Dong Kim, Jung-Jae Kim, Tomoko Ohta, Sampo Pyysalo, and Pierre Zweigenbaum. 2013. Overview of BioNLP shared task 2013. In *Proceedings of the BioNLP Shared Task 2013 Workshop*, pages 1–7.

F. Pedregosa, G. Varoquaux, A. Gramfort, V. Michel, B. Thirion, O. Grisel, M. Blondel, P. Prettenhofer, R. Weiss, V. Dubourg, J. Vanderplas, A. Passos, D. Cournapeau, M. Brucher, M. Perrot, and E. Duchesnay. 2011. Scikit-learn: Machine learning in Python. *Journal of Machine Learning Research*, 12:2825–2830.

A dictionary- and rule-based system for identification of bacteria and habitats in text

Helen V Cook

Novo Nordisk Foundation

Center for Protein Research,

Faculty of Health and Medical Sciences,

University of Copenhagen, Denmark

helen.cook@cpr.ku.dk

Evangelos Pafilis

Institute of Marine Biology,

Biotechnology and Aquaculture,

Hellenic Centre for Marine Research,

Crete, Greece

pafilis@hcmr.gr

Lars Juhl Jensen

Novo Nordisk Foundation

Center for Protein Research,

Faculty of Health and Medical Sciences,

University of Copenhagen, Denmark

lars.juhl.jensen@cpr.ku.dk

Abstract

The number of scientific papers published each year is growing exponentially and given the rate of this growth, automated information extraction is needed to efficiently extract information from this corpus. A critical first step in this process is to accurately recognize the names of entities in text. Previous efforts, such as SPECIES, have identified bacteria strain names, among other taxonomic groups, but have been limited to those names present in NCBI taxonomy. We have implemented a dictionary-based named entity tagger, TagIt, that is followed by a rule based expansion system to identify bacteria strain names and habitats and resolve them to the closest match possible in the NCBI taxonomy and the OntoBiotope ontology respectively. The rule based post processing steps expand acronyms, and extend strain names according to a set of rules, which captures additional aliases and strains that are not present in the dictionary. TagIt has the best performance out of three entries to BioNLP-ST BB3 cat+ner, with an overall SER of 0.628 on the independent test set.

1 Introduction

The biomedical literature is growing at an estimated 4% per year and as of 2016 there are at least 26 Million documents in PubMed (Lu, 2011). 12% of this work is never cited after 5 years and much of it might not reach its intended audience, effectively limiting the value of these scientific contributions (Lariviere et al., 2008). Molecular biology databases such as UniProt address this issue by manually curating domain-specific knowl-edge and providing it in a structured form (The UniProt Consortium, 2014). Despite efforts by the metagenomics community (Lombardot et al., 2006; Reddy et al., 2015; Hoopen et al., 2016), the same attention has not been given to manual curation in microbial and molecular ecology, where a lack of samples annotated with comprehensive metadata hinders comparative and integrative studies (Yilmaz et al., 2011). Both the initial creation and subsequent ongoing maintenance of such databases require a significant investment of labour and money (Attwood et al., 2015). In order to scale up this process, we need to automate the extraction of information from text.

The BioCreative and BioNLP communities are responding to this need by organising scientific literature mining challenges that aim to advance the state of the art (Arighi et al., 2014; Bossy et al., 2015). These competitions have resulted in the development of text-mining tools focusing on specific curation tasks (Bossy et al., 2015; Wang et al., 2015), one of which is the interactive EXTRACT tool that assists curators through automated named entity recognition (NER) of organisms, tissues, diseases and environments (Pafilis et al., 2015).

The BioNLP BB3 focuses on the identification of bacteria and their habitats in text. Bacteria are ubiquitous in natural and artificial environments, and play major diverse roles in shaping ecosystems. They thrive in the most extreme habitats – under the west Antarctic ice sheet (Christner et al., 2014), in alkaline hot springs (De León et al., 2013) – and they also proliferate in the most mundane habitats – such as the human body, which contains roughly an equal number of bacterial and human cells (Sender et al., 2016). Bacteria are responsible for the majority of nitrogen fixation on the planet (Galloway et al., 2004), affect the absorption of nutrients in the human gut (Semova et al., 2012), and are responsible for the

Proceedings of the 4th BioNLP Shared Task Workshop, pages 50–55,
Berlin, Germany, August 13, 2016. ©2016 Association for Computational Linguistics

deaths of approximately 1.5 million people each year from *Mycobacterium tuberculosis* infection (WHO, 2016). Given both their beneficial and pestilential impacts, it is important to understand the habitats in which bacteria grow so that they can be managed and controlled, especially in medical environments that provide care for immunocompromised patients (Sydnor and Perl, 2011), and in food processing environments which have the potential for wide distribution of contaminated products (Brackett, 1999).

The first steps towards automatically turning unstructured text into structured information about bacteria and their habitats are i) to recognize names of bacteria and habitats in a text, and ii) to resolve these to a predefined ontology or taxonomic resource. Whereas the first step can be addressed in a variety of different ways, such as using machine learning, manually crafted rules or dictionaries, the second step clearly requires the use of a dictionary.

The SPECIES and ORGANISMS resources are purely dictionary based methods that demonstrate above 85% precision and recall on identifying cellular organisms in abstracts (Pafilis et al., 2013). Further, these tools have extremely fast run times, a necessary requirement for processing large datasets. Dictionary based methods have the advantage of always correctly normalizing a term that has been tagged, but conversely they have the disadvantage of requiring an up-to-date, comprehensive dictionary. Building such a dictionary can be a difficult manual task, but it can be aided by the use of orthographic expansion rules and stopword lists. When parsing documents from a limited domain, such as biomedical literature, the dictionary required is much smaller in scope, and building one becomes feasible, as has been demonstrated by SPECIES and ORGANISMS which have been built from NCBI Taxonomy (Sayers et al., 2009).

NCBI taxonomy is a curated classification and nomenclature resource that covers all of the organisms in the Entrez sequence database (Sayers et al., 2009). Although these resources are the most comprehensive of their kind, very new and very old strains that are lacking sequences cannot be found in the NCBI taxonomy, and neither can known strains that have been spelled with uncommon misspellings. Further, acronyms that are not defined as synonyms will also not be present, meaning that a dictionary method that naively

used only the entries in the taxonomy would miss tagging such terms.

Here we present TagIt, a tool for named entity recognition and categorization of bacteria and their habitats. It primarily uses a dictionary-based approach, the results of which are extended with pattern-matching rules that handle acronyms that are not found in the dictionary and refine match boundaries to include bacterial strain names.

2 Methods

2.1 Dictionary creation

A dictionary for bacteria terms was generated from all NCBI taxonomy entities under the bacteria superkingdom (taxid: 2) (Sayers et al., 2009). The dictionary generation process is based on that used in (Pafilis et al., 2013). Briefly, NCBI taxonomy provides alternate names for each taxonomy level, which include common names, obsolete names and other synonyms, all of which were included in our dictionary. These terms were expanded to plural forms following the English and Latin rules for pluralizing nouns, and the abbreviations of Linnaean names, such as *E. coli* for *Escherichia coli*, were generated and included in the dictionary.

A dictionary for habitat terms was generated from the OntoBiotope ontology (OBT), and the names present in the ontology were expanded to their plural forms giving 8,345 terms. The habitat dictionary was expanded via synonym transfer based on manual mappings between OBT terms and their Brenda Tissue Ontology (BTO) counterparts (Chang et al., 2015). The BTO name dictionary available in the TISSUES database facilitated this process (Santos et al., 2015). This gave an additional 121,321 habitat synonyms. For example, the term "central nervous system" (OBT:000831) was expanded to include "hippocampus" and 2748 other terms, 76 of which are particular cell lines derived from nervous system tissue.

The same synonym transfer process was applied to map OBT terms to their NCBI taxonomy counterparts under the eukaryote branch (taxid: 2759). The term duck (OBT:002200), for example, was expanded with 46 synonyms including "mallard ducks", "northern mallard", "*Anas platyrhynchos*", and so forth. Terms that existed in NCBI taxonomy but not in OBT were mapped to the most specific relevant term. For example, all 145,546 names and synonyms under the NCBI taxonomy node Metazoa (taxid: 33208) that could not be mapped to anything more specific in Onto-Biotope were mapped to "animal" (OBT:000218). This gave a total of 5,106,213 additional synonyms.

Synonym transfer was also applied to OBT and the corresponding Environments Ontology (ENVO) terms (with name information from the ENVIRONMENTS tool) for an additional 54,673 synonyms (Buttigieg et al., 2013; Pafilis et al., 2015). However, as shown later, this did not improve the systems accuracy and so was not used in the final version.

Since dictionary-based NER is prone to poor precision, especially after automatic dictionary expansion, stopword lists are used to remove matches that contribute the most to the drop in precision. Here, stopword lists were generated for both bacteria and habitat entity types by manually inspecting the most frequently identified terms when tagging the Medline corpus, and removing those terms that were likely to not refer to true positive matches. This resulted in 2381 stopwords for bacteria including words such as "unclassified", and 2592 stopwords for habitat, including words such as "scales" and "root", which can have many different meanings. The full dictionaries, including the stopwords, are provided in the associated repository located at http://github.com/bitmask/BioNLP-BB3.

2.2 Tagging and post processing

Both entity types were tagged using the left-most longest matching and hashing function present in the SPECIES tool, which is case insensitive, and disregards hyphens and white space characters within names and quotes and parentheses at the beginning or end of names (Pafilis et al., 2013).

A series of post processing steps followed the tagging step. First, the input document was examined for parentheses, and these and their contents were replaced by whitespace. The tagger was run again on the modified text to identify any additional matches that spanned the parentheses. These new results were merged into the original results.

Second, the normalizations were filtered to return only the highest confidence normalization for each entry (by default SPECIES may return multiple normalizations). The normalizations for bacteria were updated so that a mention of a genus that

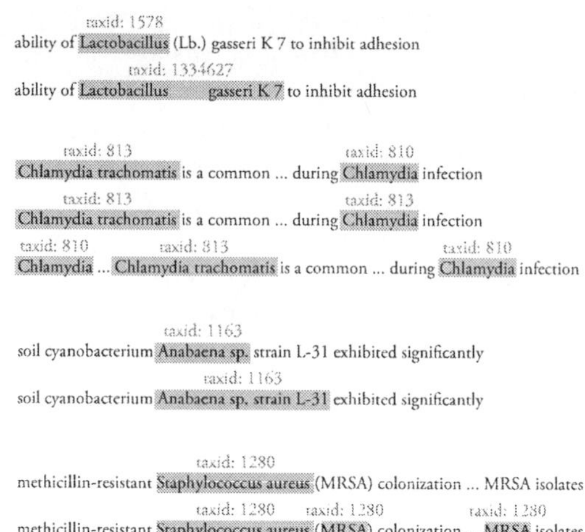

Figure 1: Illustration of the four post processing steps: Parentheses avoidance, normalization correction, strain expansion, and acronym expansion, where the first line in each block indicates the matches and normalizations prior to post processing, and the subsequent lines show how they are updated after post processing.

followed a more specific species mention (within that genus) would be normalized to the species. Although not in the annotation guidelines, we added the exception that if the genus was mentioned alone before any species within that genus, then later mentions of the genus would not be changed to refer to the specific species because such mentions were much more likely to refer to the genus in general than to have been an instance of synecdoche. These cases are illustrated in Figure 1.

Third, for bacteria, strain names were expanded by matching the text immediately following a match returned from the tagger against a regex that would identify it as a strain. Strains names were identified as sequences of letters and punctuation that may have included an indicator such as "sp." or "strain".

Lastly, acronyms were identified for both bacteria and habitats by searching the text following a match for a potential short form. Text was considered to be a short form if it was within parentheses, contained capital letters, and contained the first letter of the long form within its first three letters. Then, the remainder of the document was searched for further instances of the short form, which were normalized to the definition of the long form.

Full details and code are available at http://github.com/bitmask/BioNLP-BB3

3 Results and Discussion

Our entry, TagIt, performed best out of three entries submitted to the BioNLP-ST BB3-cat+ner task with an overall slot error rate (SER) of 0.628 on the test set. For bacteria only the SER was 0.399, and for habitats only the SER was 0.775.

TagIt uses a dictionary for both named entity recognition and for categorization, which is generated a priori from existing ontologies and rules regarding name expansion. Generating the dictionary does not require the input of any training documents, nor does this approach require that the values of any variables be learned during a training step. Therefore, we have evaluated our method on both the provided training and development sets, and see consistent performance between them.

In order to quantify the improvements from expanding the dictionaries, we generated six iterations of the dictionary that we evaluated independently on the training and development sets. The first, included only the dictionary for bacteria. The second naively added in habitats from the Onto-Biotope ontology with no synonym transfer for the habitats dictionary. The next three variants transferred synonyms to the habitats dictionary from BTO, eukaryotic entries from NCBI and ENVO, respectively. The final dictionary featured synonym transfer from both BTO and NCBI, giving better performance than either alone. This dictionary was selected as our final submission to the contest.

For both training and development sets, performance increased (i.e. SER decreased) with the addition of BTO and NCBI synonyms to the dictionary. The improvement in habitat only SER from using an unexpanded habitats dictionary, and including the mappings from BTO and NCBI – from 0.568 to 0.511 (dev) or 0.635 to 0.587 (train) – shows the performance increase possible by using synonyms in other ontologies to expand the range and number of synonyms present in the dictionary.

Adding ENVO synonyms surprisingly did not increase performance. The performance of this dictionary was evaluated in (Pafilis et al., 2015) at 87.8% precision over 600 documents, so it is unlikely that the lack of performance increase we see is due to some underlying defect in the dictionary. Further, the mapping between OBT and ENVO orthologies was performed manually by subject matter experts, so this is also unlikely to be a major source of error. The addition of ENVO synonyms cannot increase the false negative rate, as adding names to the dictionary will not result in less being found. The addition of ENVO synonyms did increase the false positive rate. The false positives included three terms that were used as homonyms such as "reservoir", intended in the dictionary to refer to a body of water, but used in the text to mean a source of bacteria. The instances of these false positives could be reduced by adding these terms to the stopword list. One further case registers as a false positive ("farms" at position 502, 507 in `BB-cat+ner-2696427.txt`), but upon manual inspection appears to be consistent with the annotation guidelines. Overall, the addition of the ENVO dictionary resulted in the identification of only a few additional terms, and if the identified errors were fixed, we would see only a minor improvement in performance compared to a dictionary without ENVO included.

In terms of the results for bacteria, the false negatives identified by TagIt included the names of strain mutants (such as Ara+), multiposition matches, and acronyms that are defined in a nonstandard manner. Bacterial false positives included a small number of cases in which terms such as "cyanobacterium" were used as adjectives or descriptions and should not have been annotated. Further, TagIt identified an additional 3 instances in which the boundaries disagreed with the gold standard, and 27 cases in which the normalizations disagreed with the gold standard, but in both cases our annotations more closely reflected the annotation guidelines.

4 Conclusions

Accurate identification of entities in text is a first necessary step towards automated extraction of information about those entities. Here, we have presented a dictionary- and rule-based system, called TagIt, to identify bacterial names and habitats which gives good performance on both entity types.

Dictionary methods for named entity recognition and categorization can give very good performance on limited domains, and rule based post processing can help overcome the intrinsic limitations to the dictionary approach. To recognize bacterial entities, applying simple rules to expand

	Overall SER			Bacteria only SER			Habitats only SER		
	Train	Dev	Test	Train	Dev	Test	Train	Dev	Test
Bacteria	0.778	0.757		0.341	0.303		n/a	n/a	
Bacteria + Habitats	0.537	0.477		0.341	0.303		0.635	0.568	
Bacteria + Habitats + BTO	0.529	0.468		0.341	0.303		0.623	0.555	
Bacteria + Habitats + NCBI	0.514	0.448		0.341	0.303		0.599	0.524	
Bacteria + Habitats + ENVO	0.540	0.479		0.341	0.303		0.639	0.572	
Bacteria + Habitats + BTO + NCBI	0.506	0.439	0.628	0.341	0.303	0.399	0.587	0.511	0.775

Table 1: Performance of TagIt in terms of overall, bacteria only and habitat only slot error rates for training, development and test sets over six variations of the dictionary (see text for their definitions).

strains and acronyms helped identify names that were not present in the dictionary. Dictionary synonym expansion also increases the performance of dictionary based methods, as was seen by the addition of BTO and NCBI synonyms to our habitats dictionary, boosting the performance over what was possible with no synonym expansion.

Acknowledgments

EU BON (EU FP7 Contract No. 308454 program), the Micro B3 Project (287589), the Earth System Science and Environmental Management COST Action (ES1103) and the Novo Nordisk Foundation (NNF14CC0001).

References

Cecilia N Arighi, Cathy H Wu, Kevin B Cohen, Lynette Hirschman, Martin Krallinger, Alfonso Valencia, Zhiyong Lu, John W Wilbur, and Thomas C Wiegers. 2014. BioCreative-IV Virtual Issue. *Database*, 2014:1–6.

Teresa Attwood, Bora Agit, and Lynda Ellis. 2015. Longevity of Biological Databases. *EMBnet.journal*, 21(0).

Robert Bossy, Wiktoria Golik, Zorana Ratkovic, Dialekti Valsamou, Philippe Bessières, and Claire Nédellec. 2015. Overview of the Gene Regulation Network and the Bacteria Biotope Tasks in BioNLP'13 Shared Task. *BMC Bioinformatics*, 16(Suppl 10):S1.

Robert E. Brackett. 1999. Incidence, Contributing Factors, and Control of Bacterial Pathogens in Produce. *Postharvest Biology and Technology*, 15(3):305–311.

Pier Luigi Buttigieg, Norman Morrison, Barry Smith, Christopher J Mungall, and Suzanna E Lewis. 2013. The Environment Ontology: Contextualising Biological and Biomedical Entities. *Journal of Biomedical Semantics*, 4(43).

Antje Chang, Ida Schomburg, Sandra Placzek, Lisa Jeske, Marcus Ulbrich, Mei Xiao, Christoph W. Sensen, and Dietmar Schomburg. 2015. BRENDA in 2015: Exciting developments in its 25th Year of Existence. *Nucleic Acids Research*, 43(D1):D439–D446.

Brent C. Christner, John C. Priscu, Amanda M. Achberger, Carlo Barbante, Sasha P. Carter, Knut Christianson, Alexander B. Michaud, Jill A. Mikucki, Andrew C. Mitchell, Mark L. Skidmore, Trista J. Vick-Majors, and the WISSARD Science Team. 2014. A Microbial Ecosystem Beneath the West Antarctic Ice Sheet. *Nature*, 512(7514):310–313.

Kara Bowen De León, Robin Gerlach, Brent M. Peyton, and Matthew W. Fields. 2013. Archaeal and Bacterial Communities in Three Alkaline Hot Springs in Heart Lake Geyser Basin, Yellowstone National Park. *Frontiers in Microbiology*, 4(330):1–10.

J. N. Galloway, F. J. Dentener, D. G. Capone, E. W. Boyer, R. W. Howarth, S. P. Seitzinger, G. P. Asner, C. C. Cleveland, P. A. Green, E. A. Holland, D. M. Karl, A. F. Michaels, J. H. Porter, A. R. Townsend, and C. J. Vörösmarty. 2004. *Nitrogen Cycles: Past, Present, and Future*, volume 70. Kluwer Academic Publishers.

Ten Hoopen, Amid C, Luigi Buttigieg, Pafilis E, Bravakos P, Cerdeño-Tárraga AM, Gibson R, Kahlke T, Legaki A, Narayana Murthy, Papastefanou G, Pereira E, Rossello M, Luisa Toribio, and Cochrane G. 2016. Value, but High Costs in Post-Deposition Data Curation. *Database*, pages 1–10.

Vincent Lariviere, Yves Gingras, and Eric Archambault. 2008. The Decline in the Concentration of Citations, 1900-2007. *Pre-print*, (arXiv:0809.5250 [physics.soc-ph]):1–9.

Thierry Lombardot, Renzo Kottmann, Hauke Pfeffer, Michael Richter, Hanno Teeling, Christian Quast, and Frank Oliver Glöckner. 2006. Megx.net– Database Resources for Marine Ecological Genomics. *Nucleic Acids Research*, 34:D390–D393.

Zhiyong Lu. 2011. PubMed and Beyond: A Survey of Web Tools for Searching Biomedical Literature. *Database*, 2011:1–13.

Evangelos Pafilis, Sune P. Frankild, Lucia Fanini, Sarah Faulwetter, Christina Pavloudi, Aikaterini

Vasileiadou, Christos Arvanitidis, and Lars Juhl Jensen. 2013. The SPECIES and ORGANISMS Resources for Fast and Accurate Identification of Taxonomic Names in Text. *PLoS ONE*, 8(6):2–7.

Evangelos Pafilis, Sune P Frankild, Julia Schnetzer, Lucia Fanini, Sarah Faulwetter, Christina Pavloudi, Katerina Vasileiadou, Patrick Leary, Jennifer Hammock, Katja Schulz, Cynthia Sims Parr, Christos Arvanitidis, and Lars Juhl Jensen. 2015. ENVIRONMENTS and EOL: Identification of Environment Ontology Terms in Text and the Annotation of the Encyclopedia of Life. *Bioinformatics*, 31(11):1872–1874.

Tatiparthi B. K. Reddy, Alex D. Thomas, Dimitri Stamatis, Jon Bertsch, Michelle Isbandi, Jakob Jansson, Jyothi Mallajosyula, Ioanna Pagani, Elizabeth A. Lobos, and Nikos C. Kyrpides. 2015. The Genomes OnLine Database (GOLD) v.5: A Metadata Management System Based on a Four Level (Meta)Genome Project Classification. *Nucleic Acids Research*, 43(D1):D1099–D1106.

Alberto Santos, Kalliopi Tsafou, Christian Stolte, Sune Pletscher-Frankild, Seán I O'Donoghue, and Lars Juhl Jensen. 2015. Comprehensive Comparison of Large-Scale Tissue Expression Datasets. *PeerJ*, 3:e1054.

Eric W Sayers, Tanya Barrett, Dennis A Benson, Stephen H Bryant, Kathi Canese, Vyacheslav Chetvernin, Deanna M Church, Michael DiCuccio, Ron Edgar, Scott Federhen, Michael Feolo, Lewis Y Geer, Wolfgang Helmberg, Yuri Kapustin, David Landsman, David J Lipman, Thomas L Madden, Donna R Maglott, Vadim Miller, Ilene Mizrachi, James Ostell, Kim D Pruitt, Gregory D Schuler, Edwin Sequeira, Stephen T Sherry, Martin Shumway, Karl Sirotkin, Alexandre Souvorov, Grigory Starchenko, Tatiana A Tatusova, Lukas Wagner, Eugene Yaschenko, and Jian Ye. 2009. Database Resources of the National Center for Biotechnology Information. *Nucleic Acids Research*, 37:D5–15.

Ivana Semova, Juliana D. Carten, Jesse Stombaugh, Lantz C. MacKey, Rob Knight, Steven A. Farber, and John F. Rawls. 2012. Microbiota Regulate Intestinal Absorption and Metabolism of Fatty Acids in the Zebrafish. *Cell Host and Microbe*, 12(3):277–288.

Ron Sender, Shai Fuchs, and Ron Milo. 2016. Revised Estimates for the Number of Human and Bacteria Cells in the Body. *bioRxiv pre-print*, pages 1–21.

Emily R M Sydnor and Trish M. Perl. 2011. Hospital Epidemiology and Infection Control in Acute-Care Settings. *Clinical Microbiology Reviews*, 24(1):141–173.

The UniProt Consortium. 2014. UniProt: A Hub for Protein Information. *Nucleic Acids Research*, 43(D1):D204–212.

Qinghua Wang, Shabbir Syed Abdul, Lara Almeida, Sophia Ananiadou, Yalbi Itzel Balderas-Martínez, Riza BatistaNavarro, David Campos, Lucy Chilton, Hui-Jou Chou, Gabriela Contreras, Laurel Cooper, Hong-Jie Dai, Juliane Fluck, Socorro Gama, Georgios Gkoutos, Afroza Khanam Irin, Lars Juhl Jensen, Silvia Jimenez, Toni Rose Jue, Ingrid Keseler, Sumit Madan, Sérgio Matos, Peter McQuilton, Matthew Mort, Jeyakumar Natarajan, Evangelos Pafilis, Emiliano Pereira, Shruti Rao, Fabio Rinaldi, David Salgado, Onkar Singh, Raymund Stefancsik, Chu-Hsien Su, Suresh Subramani, Hamsa Dhwani Tadepally, Loukia Tsaprouni, Nicole Vasilevsky, Xiaodong Wang, Andrew Chatraryamontri, Stan Laulederkind, Sherri Matis-Mitchell, Johanna McEntyre, Sandra Orchard, Sangya Pundir, Raul Rodriguez-Esteban, Kimberly Van Auken, Zhiyong Lu, Mary Schaeffer, Lynette Hirschman, and Cecilia Arighi. 2015. Overview of the Interactive Task in BioCreative V. *Proceedings of the Fifth BioCreative Challenge Evaluation Workshop*, page 20.

WHO. 2016. Tuberculosis.

Pelin Yilmaz, Jack A. Gilbert, Rob Knight, Linda Amaral-Zettler, Ilene Karsch-Mizrachi, Guy Cochrane, Yasukazu Nakamura, Susanna-Assunta Sansone, Frank Oliver Gloeckner, and Dawn Field. 2011. The Genomic Standards Consortium: Bringing Standards to Life for Microbial Ecology. *The ISME Journal*, pages 1565–1567.

A   Supplemental Material

Code, dictionaries, and the mapping of BTO and NCBI taxonomy to OBT is available at: http://github.com/bitmask/BioNLP-BB3

Ontology-based Categorization of Bacteria and Habitat Entities using Information Retrieval Techniques

Mert Tiftikci*, Hakan Şahin*, Berfu Büyüköz*, Alper Yayıkçı*, Arzucan Özgür
Department of Computer Engineering, Boğaziçi University, İstanbul, Turkey
{mert.tiftikci,hakan.sahin1,berfu.buyukoz,alper.yayikci,arzucan.ozgur}@boun.edu.tr

Abstract

A database which provides information about bacteria and their habitats in a comprehensive and normalized way is crucial for applied microbiology studies. Having this information spread through textual resources such as scientific articles and web pages leads to a need for automatically detecting bacteria and habitat entities in text, semantically tagging them using ontologies, and finally extracting the events among them. These are the challenges set forth by the Bacteria Biotopes Task of the BioNLP Shared Task 2016. This paper describes a system for habitat and bacteria entity normalization through the OntoBiotope ontology and the NCBI taxonomy, respectively. The system, which obtained promising results on the shared task data set, utilizes basic information retrieval techniques.

1 Introduction

Retrieving useful information from text became increasingly important as numerous data are collected on the Internet (Singhal, 2001). It became even more crucial to be able to reach the desired information from among lots of articles and resources when it comes to studies of science, especially biomedicine (Cohen and Hersh, 2005). The problem tackled in this paper is the semantic categorization of bacteria and habitat entities extracted from scientific paper abstracts. This problem has been addressed as a sub-task of the BioNLP Bacteria Biotope Shared Task 2016 (Deléger et al., 2016).

The Bacteria Biotope Task of the BioNLP Shared Task was previously conducted in 2011 (Bossy et al., 2011; Bossy et al., 2012) and 2013 (Bossy et al., 2013; Bossy et al., 2015). Both machine learning based (Nguyen and Tsuruoka, 2011; Björne et al., 2012; Grouin, 2013; Claveau, 2013) and rule based approaches (Ratkovic et al., 2012; Karadeniz and Ozgür, 2013; Bannour et al., 2013) have been developed to identify and normalize bacteria and habitat entities. The normalization of habitat entities through the OntoBiotope ontology has been first addressed in the 2013 edition of the Entity Categorization sub-task, where four teams participated (Bossy et al., 2013; Bossy et al., 2015). The highest F1-score (61%) and lowest Slot Error Rate (SER) (66%) was achieved by the *LIPN* system (Bannour et al., 2013), which used a combination of an ontology projection method and a rule based machine learning algorithm, namely WHISK (Soderland, 1999). The *BOUN* system (Karadeniz and Ozgür, 2013; Karadeniz and Özgür, 2015), which is based on syntactic rules, and the *LIMSI* system (Grouin, 2013), which is based on Conditional Random Fields (CRF), obtained similar SER scores (68%). The *IRISA* system, which obtained a SER value of 93% (Claveau, 2013), used the k nearest neighbor algorithm with the Okapi-BM25 (Robertson et al., 1999) similarity measure.

In this paper we describe the system that we developed for our participation at the "Entity categorization" sub-task of the Bacteria Biotope (BB3) task of BioNLP Shared Task 2016. Motivated by the promising results of rule-based entity categorization approaches in the previous editions of the

---

*These authors contributed equally to this work.

Proceedings of the 4th BioNLP Shared Task Workshop, pages 56–63,
Berlin, Germany, August 13, 2016. ©2016 Association for Computational Linguistics

shared task, we designed a rule-based approach that makes use of information retrieval and pattern matching techniques for normalizing bacteria and habitat entities through the provided ontologies.

2 System Description

2.1 Overview of the System

We developed an ontology based categorization system for the Entity Categorization sub-task. The system consists of two modules, one for the habitat categorization task and the other for the bacteria categorization task. The habitat categorization module makes use of basic information retrieval techniques including tf-idf scoring and cosine similarity. The bacteria categorization module utilizes string matching methods such as Levenshtein distance. These modules are described in detail in the following sub-sections.

2.2 Categorization of Habitat Entities

The workflow of the habitat categorization module is presented in Figure 1. Given a habitat entity mention, the goal is to identify the corresponding concepts in the OntoBiotope ontology. First, the OntoBiotope ontology is expanded by using the training and development data sets. Next, both exact matching and partial matching approaches are used to identify the ontology concepts relevant to the habitat entity mention. Partial matching is formulated as an information retrieval task, where tf-idf scoring and cosine similarity are used to rank the ontology concepts with respect to the given habitat entity.

2.2.1 Ontology Expansion

The OntoBiotope ontology is an ontology of biotopes organized as a hierarchical structure of concepts. A sample concept in the ontology is shown in Figure 2. An OntoBiotope concept consists of an ID, name, as well as exact and related synonyms. The parent-child relations between concepts are represented with the is_a field.

```
[Term]
id: OBT:000218
name: animal
synonym: "animal host" RELATED []
synonym: "animal-associated habitat" EXACT []
synonym: "animal species" RELATED []
is_a: OBT:000036 ! eukaryote host
```

Figure 2: A sample OntoBiotope ontology concept

The documents in the training and development data sets have habitat mentions labeled with their corresponding OntoBiotope concepts. We expanded the OntoBiotope ontology by including these habitat mentions as *related synonyms* to the associated concepts. Figure 3 shows the expanded version of the "animal" concept in Figure 2, where the concept has been expanded by adding the "animals" and "animal models" as related synonyms.

```
[Term]
id: OBT:000218
name: animal
synonym:"animal host" RELATED []
synonym:"animal-associated habitat" EXACT []
synonym: "animal species" RELATED []
synonym: "animals" RELATED []
synonym: "animal models" RELATED []
is_a: OBT:000036 ! eukaryote host
```

Figure 3: A sample expanded OntoBiotope ontology concept

2.2.2 Normalization

A habitat entity mention is normalized by matching it with one or more concepts in the OntoBiotope ontology. We used exact and partial matching approaches for this task.

Given a habitat entity mention, first the system searches for exact matches with the *names* or *exact synonyms* of the ontology concepts. If an exact match is found, the habitat entity is labeled with the corresponding ontology concepts.

If an exact match is not found, partial matching is performed using information retrieval techniques. Each concept in the ontology is treated as a document and an inverted index of concepts is created. The unigrams and bigrams in the names, exact synonyms, and related synonyms of concepts are represented as tf-idf weighted terms in the inverted index. The habitat entity mention is treated

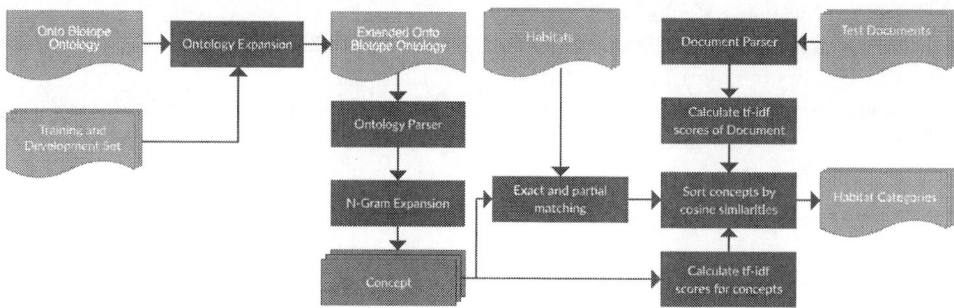

Figure 1: Categorization of Habitat Entities

as a query. In order to capture more contextual information, the query is expanded by including the unigrams and bigrams of the document where the habitat mention occurs. The cosine similarities between the query and the concepts in the inverted index are computed. The concepts are ranked based on their cosine similarity scores to the habitat query and the habitat mention is annotated with the concept that obtains the highest cosine similarity score.

If the system does not find any relevant concepts based on exact and partial matching, the habitat mention is normalized with the root of the ontology, i.e., with the concept OBT:000001 shown in Figure 4.

[Term]
id: OBT:000001
name: experimental medium
is_a: OBT:000000 ! bacteria habitat

Figure 4: Default normalization

2.3 Categorization of Bacteria Entities

The workflow of the bacteria categorization module is presented in Figure 5. In this module, bacteria entity mentions are normalized with their corresponding taxonomy IDs in the NCBI taxonomy. First, a preprocessing step is applied where punctuation marks are removed and abbreviation and acronyms are expanded. Next, a normalization and matching step is applied where preprocessed bacteria mentions are matched against the NCBI taxonomy using exact and approximate string matching methods.

2.3.1 Preprocessing

In order to increase the possibility of matching bacteria mentions with their correct categories in the NCBI taxonomy, a set of preprocessing techniques described below are developed by examining the documents and the NCBI taxonomy.

2.3.1.1 Punctuation Mark Removal

Some punctuation marks provide no useful information for our task and may hinder the performance of the system for matching bacteria names in the NCBI taxonomy. Therefore, we replaced parenthesis, quotation marks, and multiple white spaces with a single white space character. In addition, lower casing all characters is performed in this step.

The preprocessing steps described below make use of the context information, i.e., the document where the bacteria entity occurs to transform the bacteria entity mention to a more convenient format for matching with the NCBI taxonomy.

2.3.1.2 Abbreviation Expansion

One of the most common challenges for bacteria categorization is that bacteria names frequently occur in abbreviated forms. In general, the first occurrence of a bacteria name in a document is written as a full name (e.g., *"Escherichia coli"*) and the successive mentions are written in abbreviated forms (e.g., *"E. Coli"*). A bacteria mention in abbreviated form is compared with the previous closest[1] bacteria mentions in the document. If a previously occurring bacteria mention starts with the same capital letter as the abbreviated form and

---

[1]Distance between two bacteria mentions is computed based on the positions of the mentions in the document.

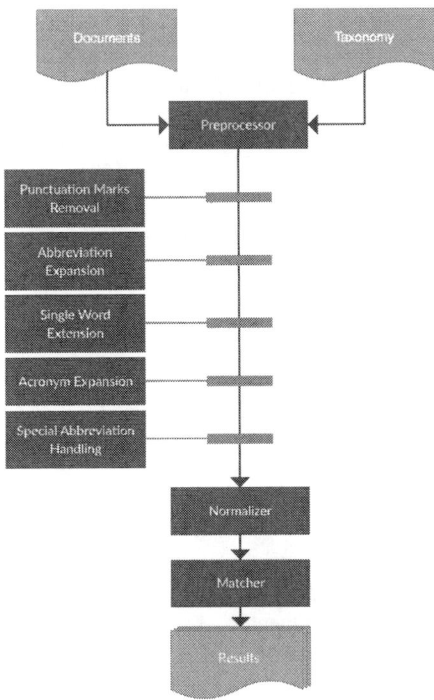

Figure 5: Categorization of Bacteria Entities

contains the remaining sub-string of the abbreviated form, the abbreviated form is converted to the corresponding bacteria mention in full name form before searching in the NCBI Taxonomy. For example, *"E. Coli"* is expanded to *"Escherichia coli"* if there is an occurrence of *"Escherichia coli"* before the abbreviated form in the same document. If there is not a match with previously occurring bacteria mentions in the document, then a search is performed starting from the abbreviated form until the end of the document, to look for the expanded version.

Another commonly occurring abbreviation pattern in documents is that the first terms of two bacteria mentions actually refer to the same word, but one of them occurs as an abbreviation. For example, in the *"Chlamydia trachomatis and C. psittaci"* phrase, *"C."* corresponds to *"Chlamydia"*. The bacteria mentions occurring before and after the abbreviated name in the document are examined. If there is not a bacteria mention in the document matching the sub-string *"psittaci"*, then bacteria mentions starting with the same letter are considered as matches. Search is performed from the abbreviated mention first to the beginning of the document and next to the end of the document. Preference is given to matches that are closer to the abbreviated form in the document. In the provided example, *"C. psittaci"* is expanded to *"Chlamydia psittaci"* before searching the NCBI Taxonomy.

2.3.1.3 Single Word Abbreviation Expansion

Another common abbreviation pattern is when after the full name mention of a bacterium, it is referred to with the first word in its name (i.e., its genus name) in the rest of the document. For example, *"Escherichia coli"* is referred to as *"Escherichia"*. To expand such abbreviations, the single word bacteria mention is compared with the preceding and following bacteria mentions in the document. If the single word bacteria mention is a sub-string of a multi-word bacteria mention, the single word mention is expanded to that multi-word mention. Preceding and closer matches are given higher precedence.

2.3.1.4 Acronym Expansion

Although bacteria entities can be referred to with acronyms like "MRSA" in documents, such acronyms are not directly represented in the NCBI taxonomy, but may appear within the names of multiple bacteria categories with different IDs such as "Staphylococcus aureus MRSA-Lux-1" and "Staphylococcus aureus MRSA-Lux-2". Based on our observation in the training set, in order to resolve ambiguity, we expanded such acronyms consisting of less than five capital letters to the closest bacteria mention in the same document.

2.3.1.5 Handling Other Special Abbreviations

Several special abbreviations including "sp.", "spp.", "strain", "str.", "aff.", "cf.", "subgen.", "gen.", and "nov." are used within species names in biomedical documents. These special abbreviations should be ignored while matching species names against the NCBI Taxonomy, since they are not in general included in the "scientific name" or the name tagged as "authority" in the NCBI Taxonomy. For instance, *"Escherichia (sp.) coli"* in a biomedical document should match with *"Escherichia coli"* in the NCBI Taxonomy. Therefore, we removed such abbreviations from bacteria name mentions in text to improve matching performance.

Another challenge in bacteria categorization is that a bacteria species can have numerous sub-types, each corresponding to a different category in the taxonomy. This makes it hard to match a bacteria mention in text with its corresponding category in the taxonomy. Special rules are designed by analyzing the provided training and development data to enhance matching in such cases. For example, the word "type" is removed from a bacteria mention in text before matching against the NCBI Taxonomy. This enables matching *"Escherichia coli type a"* in text with *"Escherichia coli a"* in the taxonomy. In cases where sub-types are denoted with semi-colon, the sub-string following the semi-colon in the bacteria mention is removed before matching with the terms in the taxonomy. This allows *"Escherichia Coli O8:K88"* in text to match with the category *"Escherichia Coli O8"* in the taxonomy. Other transformations performed to enhance sub-type matching are converting the "ssp" abbreviation to "subsp." and the "ara+" sub-string to "ara+ biotype" in the bacteria mentions in text. We did not remove these sub-species denoting abbreviations, since keeping them resulted in better performance. We converted these abbreviations to their versions occurring in the names tagged as "scientific name" or "authority" in the taxonomy.

2.3.2 Normalization and Matching with Taxonomy

After the preprocessing steps, a bacteria mention in text is converted to a candidate phrase to be matched against the categories in the NCBI Taxonomy. First, an exact match is performed and the phrase is assigned to the matching category in the taxonomy.

If there is no an exact match, then partial phrase matching is performed. In a candidate phrase, it is possible that an irrelevant word, for instance an adjective, appears. That irrelevant word will cause an unsuccessful search in the taxonomy. Therefore, partial matching with the first two words, last two words, and first and last words of the candidate phrase are performed and the partially matching category is assigned to the candidate phrase.

If exact and partial phrase matching do not match with any categories in the taxonomy, then partial string matching using *Levenshtein edit distance* is performed to detect the most similar cat-

egory to the candidate. We set the edit distance threshold to 2. For taxonomy categories with edit distance less than or equal to the threshold, "error ratio" is computed as follows.

$$\textbf{Error ratio} = \frac{edit\ distance}{length\ of\ the\ candidate} \quad (1)$$

The error ratio threshold is set to 0.2. So, the candidate phrase is assigned to a taxonomy category, if edit distance and error ratio are less than or equal to 2 and 0.2, respectively. In this case, if a candidate phrase is of length 4, and if a bacteria name is found in the taxonomy with edit distance 1, this is not accepted as a successful match, since error ratio is 0.25.

Finally, if an exactly or partially matching category is not found, the context of the bacteria mention in the document is used for category assignment. In this case, the bacteria mention is mapped to the same category of the closest bacteria mention for which a category was assigned in the document.

3 Evaluation and Results

Different evaluation metrics are used for habitat and bacteria entities. Wang similarity (Wang et al., 2007) with a weight of 0.65 is used for evaluation of habitat entities by computing the similarity between the reference and the predicted normalization. This metric determines a semantic similarity score between two nodes of a directed acyclic graph (DAG) whose nodes have is_a relations with their parents. This similarity metric takes into account the locations of the terms within the DAG, their distances to the root and to common ancestors. Evaluation of bacteria entities is stricter. If two terms (reference and predicted) are the same, the similarity score is equal to 1, otherwise it is 0. Two systems, namely LIMSI and our system BOUN participated in the BB3-CAT shared task. The official evaluation results on the shared task test data set are presented in Table 1[2]. Among the two participating systems, our system ranked first in the overall task of habitat and bacteria name categorization, as well as in the individual sub-tasks of habitat name categorization and bacteria name categorization.

Precision	BOUN	LIMSI
Main Scoring	0.679	0.503
Habitats Only	0.620	0.438
Bacteria Only	0.801	0.637

Table 1: Official evaluation results

3.1 Results for Habitat Categorization

This sub-section provides the evaluation results of the system at its major development phases over the training and development data sets. Precision and recall values calculated from true positives, false positives, and false negatives are reported. Habitats that are normalized correctly are considered to be true positive, the ones normalized with wrong categories are considered to be false positives, and if there are no exact or partial matches found for a habitat, it is considered to be a false negative. BB3-CAT Precision corresponds to the entity categorization precision computed using the online evaluation tool provided at the BB3 shared task[3]. BB3-CAT precision is based on the Wang similarity (Wang et al., 2007).

	Development	Training
True Positive	214	309
False Positive	186	349
False Negative	321	516
Precision	0.53	0.47
Recall	0.40	0.37
BB3-CAT Precision	0.58	0.55

Table 2: Results without bigram expansion or normalization to OBT:000001

Table 2 presents the baseline results when only unigrams are used for cosine similarity computation and no normalization to the root concept is performed. Table 3 presents the results after introducing the bigrams to the system and simultaneously increasing the term frequency weights of the unigrams by a factor of two.

	Development	Training
True Positive	224	332
False Positive	176	326
False Negative	311	493
Precision	0.56	0.50
Recall	0.42	0.40
BB3-CAT Precision	0.61	0.59

Table 3: Results with bigram expansion

Although bigram expansion increases the scores, there are still some habitats with no matched categories. In case of computing Wang scores, leaving a habitat without a category is a drawback, since any normalization gains a better score than no normalization. Our results in Table 4 show that normalizing unmatched habitats to the concept OBT:000001 increases the Wang scores.

	Development	Training
True Positive	226	343
False Positive	228	404
False Negative	309	482
Precision	0.50	0.46
Recall	0.42	0.42
BB3-CAT Precision	0.63	0.62

Table 4: Results with bigram expansion and normalization to OBT:000001

3.2 Results for Bacteria Categorization

Our baseline system that only performs punctuation removal and exact matching between candidate name and a bacteria name in the taxonomy obtained precision-F-measure values of 0.39-0.40 over the development set and 0.57-0.58 over the training set. This benchmark was a plain starting point for this study.

We improved the baseline system by applying the preprocessing steps. The most common errors seen in the results were the unmatched abbreviations. Then, we applied the abbreviation expansion step described in Sub-section 2.3.1.2. and our precision-F-measure values increased to 0.59-0.71 over the development set and to 0.74-0.83 over the training set. Thus, this enhancement was the most effective one overall.

After that, we applied the single word abbreviation expansion step described in Sub-section 2.3.1.3. This improvement increased the precision-F-measure values to 0.64-0.75 on the de-

---

[3]http://bibliome.jouy.inra.fr/demo/BioNLP-ST-2016-Evaluation/index.html

velopment set and to 0.78-0.86 on the training set. Finally, we applied the acronym expansion step and it raised the precision-F-measure values to 0.67-0.77 on the development set and to 0.81-0.87 on the training set. This final result is the last benchmark that we got after preprocessing the bacteria names. The results of the preprocessing steps are presented in Table 5.

		Development	Training
Punctuation rem.	P	0.39	0.57
	R	0.41	0.59
	F	0.40	0.58
Abbreviation exp.	P	0.59	0.74
	R	0.89	0.94
	F	0.71	0.83
Single word exp.	P	0.64	0.78
	R	0.90	0.95
	F	0.75	0.86
Acronym exp.	P	0.67	0.81
	R	0.90	0.93
	F	0.77	0.87

Table 5: Results after preprocessing (P: Precision, R: Recall, F: F-measure)

Table 6 summarizes the results of the normalization and matching steps that are performed after the preprocessing steps. Matching with the original bacteria mention first and matching with the preprocessed version if there is no a match with the original version resulted in 0.77 precision and 0.78 F-measure over the development set and 0.87 precision and 0.88 F-measure over the training set. In addition, both partial phrase matching using two-word combinations and partial string matching using Levenshtein distance resulted in improved performance. Finally, assigning unmatched bacteria mentions to the taxonomy of the closest categorized bacteria mention in the same document resulted in considerable improvement in precision and F-measure.

4 Conclusion

This study introduced a system that is developed in the scope of the Entity Categorization sub-task of the BioNLP Bacteria Biotope Shared Task 2016. The system consists of two modules both of which target normalizing entities that have been detected in scientific paper abstracts. While the habitat categorization module operates on habitat mentions expressed in natural language and uses the OntoBiotope ontology for normalization, the bacteria categorization module deals with bacteria mentions expressed as more structured scientific ex-

		Development	Training
Exact matching	P	0.77	0.87
	R	0.79	0.89
	F	0.78	0.88
Sub-phrases	P	0.81	0.90
	R	0.85	0.92
	F	0.83	0.91
Edit distance	P	0.83	0.91
	R	0.89	0.95
	F	0.86	0.93
Unmatched handling	P	0.89	0.95
	R	0.99	0.99
	F	0.94	0.97

Table 6: Results after the matching and normalization steps (P: Precision, R: Recall, F: F-measure)

pressions and uses the NCBI Taxonomy for normalization.

Promising results are obtained by both modules, which utilize pattern matching and information retrieval techniques. According to the official evaluations, the habitat categorization module obtained 0.620 precision and the bacteria categorization module obtained 0.801 precision, which led to achieving the highest overall precision of 0.679 in the BB3-CAT sub-task. As future work, integrating WordNet based similarity measures to improve ontology-based matching will be investigated.

Acknowledgments

This work has been supported by Marie Curie FP7-Reintegration-Grants within the 7th European Community Framework Programme. We would like to thank the Bacteria Biotope Task organizers for organizing the shared task and for their help with the questions.

References

Sondes Bannour, Laurent Audibert, and Henry Soldano. 2013. Ontology-based Semantic Annotation: An Automatic Hybrid Rule-based Method. In *Proceedings of the BioNLP Shared Task 2013 Workshop*, pages 139–143. Sofia August.

Jari Björne, Filip Ginter, and Tapio Salakoski. 2012. University of Turku in the BioNLP'11 Shared Task. *BMC Bioinformatics*, 13(11):1.

Robert Bossy, Julien Jourde, Philippe Bessieres, Maarten Van De Guchte, and Claire Nédellec. 2011. BioNLP Shared Task 2011: Bacteria Biotope. In *Proceedings of the BioNLP Shared Task 2011 Workshop*, pages 56–64. Association for Computational Linguistics.

Robert Bossy, Julien Jourde, Alain-Pierre Manine, Philippe Veber, Erick Alphonse, Maarten Van

De Guchte, Philippe Bessières, and Claire Nédellec. 2012. BioNLP Shared Task-The Bacteria Track. *BMC Bioinformatics*, 13(11):1.

Robert Bossy, Wiktoria Golik, Zorana Ratkovic, Philippe Bessières, and Claire Nédellec. 2013. BioNLP Shared Task 2013–An Overview of the Bacteria Biotope Task. In *Proceedings of the BioNLP Shared Task 2013 Workshop*, pages 161–169.

Robert Bossy, Wiktoria Golik, Zorana Ratkovic, Dialekti Valsamou, Philippe Bessières, and Claire Nédellec. 2015. Overview of the Gene Regulation Network and the Bacteria Biotope Tasks in BioNLP'13 Shared Task. *BMC Bioinformatics*, 16(Suppl 10):S1.

Vincent Claveau. 2013. IRISA Participation to Bionlp-ST 2013: Lazy-Learning and Information Retrieval for Information Extraction tasks. In *BioNLP Workshop, Colocated with ACL 2013*, pages 188–196.

Aaron M Cohen and William R Hersh. 2005. A Survey of Current Work in Biomedical Text Mining. *Briefings in Bioinformatics*, 6(1):57–71.

Louise Deléger, Robert Bossy, Estelle Chaix, Mouhamadou Ba, Arnaud Ferré, Philippe Bessières, and Claire Nédellec. 2016. Overview of the Bacteria Biotope Task at Bionlp Shared Task 2016. In *Proceedings of the 4th BioNLP Shared Task Workshop*, Berlin, Germany, August. Association for Computational Linguistics.

Cyril Grouin. 2013. Building a Contrasting Taxa Extractor for Relation Identification from Assertions: Biological Taxonomy & Ontology Phrase Extraction System. *ACL 2013*, 144.

Ilknur Karadeniz and Arzucan Ozgür. 2013. Bacteria Biotope Detection, Ontology-based Normalization, and Relation Extraction using Syntactic Rules. In *Proceedings of the BioNLP Shared Task 2013 Workshop*, pages 170–177.

İlknur Karadeniz and Arzucan Özgür. 2015. Detection and Categorization of Bacteria Habitats using Shallow Linguistic Analysis. *BMC Bioinformatics*, 16(Suppl 10):S5.

Nhung TH Nguyen and Yoshimasa Tsuruoka. 2011. Extracting Bacteria Biotopes with Semi-Supervised Named Entity Recognition and Coreference Resolution. In *Proceedings of the BioNLP Shared Task 2011 Workshop*, pages 94–101. Association for Computational Linguistics.

Zorana Ratkovic, Wiktoria Golik, and Pierre Warnier. 2012. Event Extraction of Bacteria Biotopes: A Knowledge-Intensive NLP-based Approach. *BMC Bioinformatics*, 13(11):1.

Stephen E Robertson, Steve Walker, Micheline Beaulieu, and Peter Willett. 1999. Okapi at TREC-7: Automatic Ad Hoc, Filtering, VLC and Interactive Track. *Nist Special Publication SP*, pages 253–264.

Amit Singhal. 2001. Modern Information Retrieval: A Brief Overview. *IEEE Data Eng. Bull.*, 24(4):35–43.

Stephen Soderland. 1999. Learning Information Extraction Rules for Semi-Structured and Free Text. *Machine Learning*, 34(1-3):233–272.

James Z Wang, Zhidian Du, Rapeeporn Payattakool, S Yu Philip, and Chin-Fu Chen. 2007. A New Method to Measure the Semantic Similarity of GO Terms. *Bioinformatics*, 23(10):1274–1281.

Identification of Mentions and Relations between Bacteria and Biotope from PubMed Abstracts

Cyril Grouin

LIMSI, CNRS, Université Paris-Saclay

Bât 508, Campus Universitaire, F-91405 Orsay

`cyril.grouin@limsi.fr`

Abstract

This paper presents our participation in the Bacteria/Biotope track from the 2016 BioNLP Shared-Task. Our methods rely on a combination of distinct machine-learning and rule-based systems. We used CRF and post-processing rules to identify mentions of bacteria and biotopes, a rule-based approach to normalize the concepts in the ontology and the taxonomy, and SVM to identify relations between bacteria and biotopes. On the test datasets, we achieved similar results to those obtained on the development datasets: on the categorization task, precision of 0.503 (gold standard entities) and SER of 0.827 (both NER and categorization); on the event relation task, F-measure of 0.485 (gold standard entities, ranking third out of 11) and of 0.192 (both NER and event relation, ranking first); on the knowledge-based task, mean references of 0.771 (gold standard entities) and of 0.202 (both NER, categorization and event relation).

1 Introduction

In this paper, we present the methods we used while participating in the Bacteria/Biotope track from the 2016 BioNLP Shared-Task. We partially reused the method we designed while participating in the previous edition of the challenge (Grouin, 2013), and we updated afterwards while designing new experiments (Lavergne et al., 2015).

2 Background

Four teams participated in the Bacteria/Biotope track (Bossy et al., 2015) from the 2013 BioNLP Shared-Task.

On the entity detection and categorization task, the best results were obtained using either machine-learning approaches, as done by Bannour et al. (2013) who ranked first (Slot Error Rate (SER) of 0.661), or using syntactic hand-coded rules, as done by Karadeniz and Özgür (2013) who ranked second (SER=0.676). We ranked third (SER=0.678) using CRF and normalization rules.

On the localization relation extraction task, the best results were obtained through machine-learning approaches. Björne and Salakoski (2013) ranked first (F=0.42), using a system based on Support Vector Machine (SVM), while Claveau (2013) ranked second (F=0.40) using a lazy machine learning (kNN) approach.

3 Task description

3.1 Presentation

The 2016 Bacteria/Biotope track[1] (Deléger et al., 2016) consists in three main objectives: (i) named entity recognition (NER) to identify mentions of bacteria and biotopes from scientific abstracts, (ii) categorization to normalize mentions of bacteria in the NCBI taxonomy and mentions of biotopes in the OntoBiotope ontology, and (iii) event extraction to identify relations of localization between a bacteria and a biotope.

The track is organized into three main tasks, based on gold standard annotations of entities: a categorization task (cat), an event extraction task (event), and a knowledge-base population task (kb) which combines categorization and relation identification. Additionally, each task is composed of a named entity recognition sub-task: categorization and relation identification are based on predictions of entities (cat+ner, event+ner, and kb+ner tasks) instead of gold standard annotations.

---

[1] `http://2016.bionlp-st.org/`

Proceedings of the 4th BioNLP Shared Task Workshop, pages 64–72,
Berlin, Germany, August 13, 2016. ©2016 Association for Computational Linguistics

3.2 Material

3.2.1 Corpus

The corpus is composed of 215 scientific texts (title and abstract) focusing on bacteria, extracted from the Medline database. This corpus is split into three datasets: training (71 texts), development (36 texts), and test (108 texts).[2] We used the train dataset to develop our systems and to tune our models while results produced by those systems were evaluated on the dev dataset. The test datasets were used for the official evaluation.

3.2.2 Annotations

Bossy et al. (2016) defined three kinds of entities (bacteria, habitat, geographical) and one type of relation (lives in) between a bacteria and a biotope.

Entities Annotations of entities imply three kinds of annotations: (i) single entities, (ii) embedded entities, in case of different meanings, and (iii) discontinuous entities, to deal with coordination. Figure 1 highlights discontinuous annotations (*throat cultures*) and embedded annotations (*throat* within *throat cultures*, and *nasopharyngeal* within *nasopharyngeal cultures*).

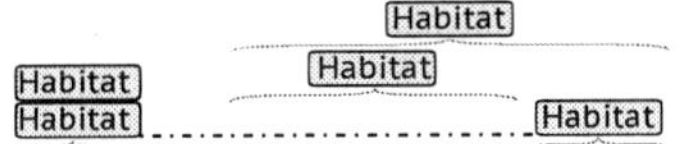

Figure 1: Discontinuous and embedded annotations of entities

Specific annotation rules apply for classifiers (*genus, species, strain*) and generic classes (*bacteria, cohort, in vivo, microbe, suspension*) which must not be annotated, except for specified strain (*mutants, serotypes, serovars*).

Categorization The categorization focuses on two types of entity (bacteria, habitat). Annotations provide the ID for each mention to be normalized, based on the NCBI taxonomy[3] (Federhen, 2002) for mentions of bacteria and the OntoBiotope ontology[4] (Nédellec, 2016) for mentions of habitat.

Mentions of bacteria are normalized into only one category while mentions of habitat can be normalized into several categories. The categorization into one or several categories for habitat mentions is dependent on the structure of the ontology, whether an "is a" relation between category candidates exists in the ontology or not (see figure 2). As an example, the mention *chicks* is normalized into three categories ("laboratory animal—000323", "infant–002177", "chicken–002229") while all mentions of *mice* are normalized into one category ("laboratory mice—002153") since this category is related with the category "laboratory animal—000323".

```
[Term]
id: OBT:000323
name: laboratory animal
is_a: OBT:000218 ! animal

[Term]
id: OBT:002153
name: laboratory mice
is_a: OBT:001865 ! mouse
is_a: OBT:000323 ! laboratory animal

[Term]
id: OBT:002229
name: chicken
is_a: OBT:002165 ! poultry
```

Figure 2: Extract from the OntoBiotope ontology

Relations Annotations of relations always imply one bacteria with one or several biotopes (habitat, geographical). Figure 3 shows relations between a bacteria and two biotopes, a geographical unit (*UK*) and a habitat (*UK retail poultry*). According to the guidelines, even if arguments from a relation must be as close as possible, one can find a few cases of relations between two distant entities. The longest distance is of 1868 characters, 276 words, implying 10 sentences.

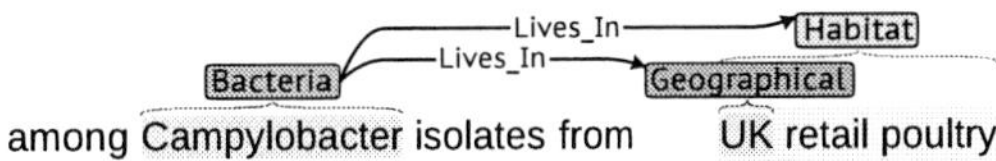

Figure 3: Example of relation annotations

3.2.3 Statistics

We present in table 1 the number of annotations for each category of entities (bacteria, habitat, geographical) and relations (lives in), as well as the number of categorizations performed in the associated resource (OntoBiotope ontology or NCBI

---

[2]The test dataset is split into two datasets: one set of 54 files for all tasks implying a named entity recognition (NER) process (cat+ner, event+ner, kb+ner) and a second set of 54 files giving gold standard annotations of bacteria and biotope for tasks without NER (cat, event, kb).

[3]http://www.ncbi.nlm.nih.gov/taxonomy

[4]http://2016.bionlp-st.org/tasks/bb2/OntoBiotope_BioNLP-ST-2016.obo

taxonomy) in each dataset (train, dev, and test[5]). The figures presented in cells with a grey background refer to the number of predictions to be made during the challenge. While annotations of entities are found in almost all files (one file from the train dataset does not propose any annotation), relations are found in about 80% of files (i.e., 84 files out of 107 files from the train+dev datasets). The number of annotated entities per file is quite unbalanced, from 1 to 69 entities.

Annotations		Train	Dev	Test #1	Test #2
Number of files		71	36	54	54
Entities	Bacteria	375	244	341	401
	Habitat	747	454	720	621
	Geographical	36	38	37	27
Category	NCBI Taxon	376	245	347	401
	OntoBiotope	825	535	861	681
Relations	Lives in	327	223	340	314

Table 1: Number of annotations per category in each dataset (test #1=dataset with reference annotations of entities, #2=dataset without annotations). Grey background refers to the number of predictions to be made during the challenge

We observed that discontinuous entities: (i) mainly concern habitat entities (87.0%), (ii) generally involve two entities, more rarely three entities, and that (iii) the pivot shared by discontinuous and continuous entities is generally at the end of the portion (e.g., "cultures" in *throat and nasopharyngeal cultures*). In the training and development datasets (107 files), out of 1894 annotations of entities, we only found 46 discontinuous entities (i.e., 2.4% of annotations are discontinuous entities).

4 Methods

Based on the three main objectives of the track and the previous observations, we considered distinct systems (cf. figure 4): named entity recognition, categorization, and relation identification. We did not use any of the provided supporting resources. Due to the low number of discontinuous entities, we decided not to process this type of annotation.

4.1 Additional data

Presentation In order to improve the robustness of our systems, we annotated a new set of 22 files.[6] To produce this new set, we queried PubMed with names of bacteria we randomly selected from the train and development datasets: *Francisella, Lactobacillus, LVS, Mycoplasma, Rickettsia, Trichomonas vaginalis* and *Vibro parahaemolyticus*. Among all results returned by PubMed, we kept abstracts published in 2016 we found interesting.

Annotations We used our systems (see sections 4.2 and 4.4) to automatically pre-annotate this dataset. One human annotator corrected and completed the automatic pre-annotations in one hour using the BRAT annotation tool (Stenetorp et al., 2012). Since we were not trained to annotate such files, even if we tried to follow the guidelines (Bossy et al., 2016), we hope our annotations are not too much inconsistent with annotations provided by the organizers. Our dataset includes 252 annotations of bacteria, 176 habitat, 31 geographical and 130 relations. Except for habitat and relations, this distribution is consistent with statistics presented in table 1.

4.2 Named Entity Recognition

4.2.1 Presentation

We considered the named entity recognition (NER) issue as a classification task, where tokens from a text should be classified into three categories (bacteria, habitat, geographical). Our NER system relies both on machine-learning approach and post-processing rules.

Machine-learning Conditional Random Fields (CRF) (Lafferty et al., 2001) are widely used for sequence labeling tasks. Our experiments rely on the Wapiti system (Lavergne et al., 2010), based on the linear-chain CRFs framework.

The feature sets are: (i) the token itself, (ii) token typographic case, presence of punctuation marks in the token, presence of digits in the token, token length, (iii) identification of the token in the OntoBiotope ontology or in the NCBI taxonomy, (iv) semantic class of the token among 37

---

[5]Test #1 refers to the test dataset with gold standard annotations of entities (cat+ner, event+ner, kb+ner tasks) while test #2 refers to the test dataset without annotations of entities (cat, event, kb tasks).

[6]Our additional dataset, annotated before the release of the test datasets, is composed of files (title and abstract) corresponding to the following PMIDs: 1262454, 21624472, 26358917, 26510639, 26678135, 26709916, 26773254, 26901499, 26902724, 26919818, 26941131, 26941728, 26942354, 26950451, 26951983, 26961264, 26962869, 26964722, 26965788, 26965874, 26968160, 26968657. None of our additional data are also part of the test datasets.

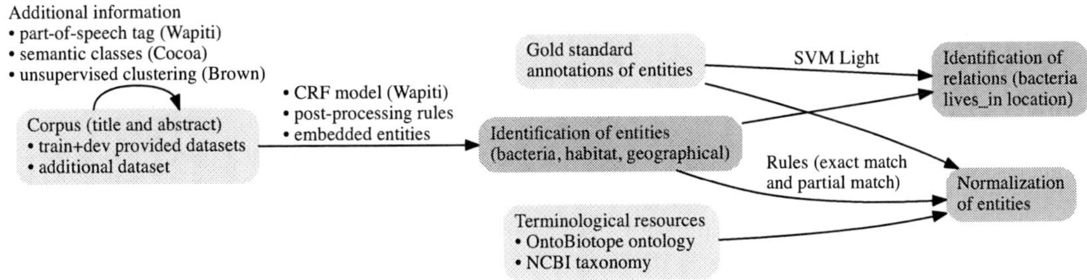

Figure 4: Systems used to identify entities, normalize entities, and identify relations

pre-defined classes *(Body part, Chemical, Food, Habitat, Organism, Physiology, etc.)*, provided by the Cocoa web API,[7] *(v)* part-of-speech tag[8] of the token, and *(vi)* cluster ID of each token through an automatic unsupervised clustering of all tokens from the train and dev datasets into 120 clusters, using the algorithm designed by Brown et al. (1992) and implemented by Liang (2005).

Since a lot of tokens from texts are not mentions of bacteria, habitat and geographical,[9] those unannotated tokens lead to an unbalanced distribution of data. This may imply an over-training of the CRF system of the unannotated tokens. In order to reduce this over-training issue, we deleted portions of unannotated tokens. Specifically, we deleted parts of text composed of unannotated tokens, if those parts are distant of more than 16 tokens[10] from the closest annotated token. As a consequence, we kept the wholeness of the context of annotated parts and we reduced the number of unannotated tokens in our training set.

We tuned our system to predict widest entities since we considered that shorter entities can easily be identified through post-processing rules. Because embedded entities only concern habitats, this strategy does not concern bacteria and geographical units. So that the CRF produces widest entities, in case of embedded annotations, we only kept the widest entities in the sample file given as input to train the CRF model.

Post-processing In order to improve the predictions we made in the previous step and to deal with some of the specific annotation rules defined in the guidelines (Bossy et al., 2016), we designed a few post-processing rules:

- annotation of abbreviations *(EHEC, EPEC, LVS, MRSA, etc.)*, generic classes with an initial upper case *(Bacteria, Bacterium)*, some nomenclatural suffixes *(sp., spp.)*, adjectives for habitat *(aquatic, nosocomial, saprophyte)* and geographical *(northern, southern, etc.)*;

- deletion of annotations for generic classes *(bacteria, bacterial, bacterium)*, modifiers *(methicillin-resistant, pathogenic)*, some nomenclatural suffixes *(gen. nov., sp. nov.)*, and 34 generic habitat terms *(antibiotic, ecosystem, world, etc.)*.

Embedded entities Since our CRF predicted widest entities, we processed embedded habitat entities through a post-processing system. For all predictions of mentions of habitat, we searched for shortened entities within widest entities. As an example, based on the prediction *gastric mucosa-associated lymphoma*, this simple rule allows us to identify the single mention *gastric*. We thus increased the coverage of the habitat mentions.

4.2.2 Design of experiments

We designed several experiments, depending on the size of the training corpus and whether we used or not post-processing rules and embedded entities processing. Results are presented in section 5.1.1. The configuration we used on the test dataset is the following one: we trained the final CRF model on all available annotated files (193 files),[11] we

---

[7]Cocoa: compact cover annotator for biological noun phrases, http://npjoint.com/annotate.php

[8]POS tagging was performed using an English POS CRF model for Wapiti: https://wapiti.limsi.fr/

[9]Based on our tokenization, among 15 530 tokens from the training dataset, only 2 110 of them (i.e., 13.59%) are part of bacteria, habitat and geographical mentions.

[10]This distance of 16 tokens has been chosen empirically. This threshold reduced by 23.1% the number of unannotated tokens in the training dataset. From now on, the 2 110 annotated tokens represent 20.45% of all tokens.

[11]Those annotated files came from the training dataset (71 files), the development dataset (36 files), the additional dataset we manually annotated (22 files), and the test #1

applied post-processing rules to correct the CRF outputs, and we processed the embedded entities through a last script.

4.3 Categorization

Exact match We performed the categorization task using a basic rule-based approach. We searched the mention to normalize in the Onto-Biotope ontology (habitat) or in the NCBI taxonomy (bacteria), through an exact match search, and returned the corresponding numeric identifier.

Partial match Additionally, we searched for partial matching of mentions of bacteria in the taxonomy: (i) shortened versions: *H. pylori* vs. *Helicobacter pylori*, (ii) specified versions: *bacillus intermedius s3-19* vs. *bacillus intermedius*, and (iii) linguistics variations: plural form (*lactobacilli* vs. *lactobacillus*) or adjectival derivation (*mycobacterial* vs. *mycobacteria*). Similarly, we searched for partial matching of mentions of habitat in the ontology: (i) linguistic variations: plural forms (*patients* vs. *patient*), hand-coded nominalization of adjectives (*clinical* vs. *clinic*), (ii) split of multi-terms into single terms (*human* and *blood* vs. *human blood*), and (iii) hand-coded transformation of specific cases (*adult* is replaced by *human adult*; *children* is replaced by *child*).

Default value At last, we defined default values for all unmatched mentions of bacteria and habitat, based on the most used values in the training and development datasets (this choice is not relevant for all unmatched mentions but it allows us to slightly improve our results). We used the taxonomy entry #210 (i.e., *Campilobacter pylori* and *Helicobacter pilori*) for bacteria, and the Onto-Biotope entry #002216 *patient with infectious disease* (the second most used category) for habitat.

4.4 Relation Extraction

In order to identify relations between bacteria and biotope, we designed experiments based on the SVM framework (Vapnik, 1995), as done by Björne and Salakoski (2013). Our experiments rely on the SVM Light implementation proposed

by Joachims (1999). Since a few long distance relations exist, in order to ensure the robustness of our system, we decided to remove all relations implying a distance higher than 80 tokens between both entities from our training set. This threshold produced the best results. It allows us to keep the shortest relations from the training dataset (i.e., 60% of all positive relations). We strictly balanced positive and negative examples to train our model.

The feature sets are: (i) a bag of words of all tokens from both entities to be linked, and (ii) the distance in characters between those entities.

5 Results

5.1 Development dataset

In this section, we present the results we achieved on the development dataset. Since we produced outputs compatible with the BRAT annotation tool, results were computed using the BRATeval evaluation tool developed by Verspoor et al. (2013) and updated by Deléger et al. (2014). This evaluation tool allows us to evaluate all kinds of entities (single, embedded and discontinuous entities) as well as relations between entities.

5.1.1 Named entity recognition

Table 2 presents the results we achieved on the development dataset in the named entity recognition sub-task. We give both the F-measure we achieved on each category (bacteria, habitat, geographical) and the detailed overall results (exact match). We designed five experiments:

1. CRF model trained on the train dataset (71 files);

2. CRF model trained on the train+additional datasets (93 files);

3. CRF model trained on the train+additional datasets (93 files) using an over-training reduction function (we reduced the number of tokens which must not be annotated);

4. CRF model trained on the train+additional datasets (93 files) using an over-training reduction function, and post-processing rules were applied (all categories);

5. CRF model trained on the train+additional datasets (93 files) using an over-training reduction function, post-processing rules were applied (all categories), and embedded entities (habitat) were processed.

---

dataset (54 files). For clarification, the named entity recognition evaluation (cat+ner, event+ner, kb+ner tasks) is performed on the test #2 dataset, composed of different files than the test #1 dataset. As a consequence, since there is no common files between test datasets #1 and #2, the use of the annotated files from the test #1 dataset to train the final CRF model does not hedge the official evaluation.

	Entity F-measures			Overall results		
#	Bact	Hab	Geo	P	R	F
1	0.668	0.470	0.727	0.721	0.452	0.556
2	0.769	0.462	0.739	0.753	0.488	0.592
3	0.772	0.469	0.739	0.740	0.500	0.597
4	0.785	0.469	0.739	0.747	0.504	0.602
5	0.785	0.523	0.739	0.737	0.548	0.628

Table 2: Results on the development dataset, F-measure for each category (Bact=Bacteria, Hab=Habitat, Geo=Geographical) and overall results (P=Precision, R=Recall, F=F-measure) depending on the experiment

5.1.2 Categorization

Table 3 presents the results we achieved on the development dataset for the categorization task. Our evaluation only computes an exact match between the IDs from the taxonomy and the ontology provided in the hypothesis and the reference. This evaluation does not compute any similarity distance within the hypothesis and reference categories. We give the overall and detailed results for both the OntoBiotope ontology and the NCBI taxonomy. Results are provided for two tasks:

1. categorization performed on the entities identified by our CRF system, configuration #5 (cat+ner task);

2. categorization performed on the gold standard annotations of entities (cat task).

#	Evaluation	P	R	F
1	Overall	0.404	0.286	0.335
	OntoBiotope	0.509	0.360	0.422
	NCBI taxonomy	0.621	0.457	0.527
2	Overall	0.456	0.412	0.433
	OntoBiotope	0.570	0.515	0.541
	NCBI taxonomy	0.886	0.885	0.886

Table 3: Results (exact match) on the development dataset on the categorization tasks (P=Precision, R=Recall, F=F-measure)

5.1.3 Relations

Table 4 presents the results we achieved (exact match) on the development dataset in the relation identification task. We designed four experiments:

1. SVM model trained on the train dataset (71 files), prediction of entities from the CRF system (event+ner task);

2. SVM model trained on the train+additional dataset (93 files), prediction of entities from the CRF system (event+ner task);

3. SVM model trained on the train dataset (71 files), gold standard annotations of entities (event task);

4. SVM model trained on the train+additional dataset (93 files), gold standard annotations of entities (event task).

#	Evaluation	P	R	F
	Entities from the CRF system (event+ner task)			
1	Overall	0.171	0.213	0.189
	Bacteria-Habitat	0.162	0.235	0.192
	Bacteria-Geographical	0.364	0.111	0.170
2	Overall	0.190	0.213	0.201
	Bacteria-Habitat	0.181	0.235	0.204
	Bacteria-Geographical	0.400	0.111	0.174
	Entities from the gold standard (event task)			
3	Overall	0.381	0.652	0.480
	Bacteria-Habitat	0.355	0.658	0.461
	Bacteria-Geographical	0.622	0.622	0.622
4	Overall	0.385	0.652	0.484
	Bacteria-Habitat	0.357	0.658	0.463
	Bacteria-Geographical	0.657	0.627	0.639

Table 4: Results on the development dataset on the relation identification tasks (P=Precision, R=Recall, F=F-measure)

5.1.4 Online evaluation service

Since the online evaluation service provides a distinct evaluation (giving final scores and using different metrics), in order to compare the results we achieved on both the development and the test datasets, we present in table 5 the results we achieved on all tasks on the development datasets using our last configuration, as computed by the evaluation service.

5.2 Test dataset (official results)

Table 6 presents the results we achieved on the test dataset. Our results are similar to results obtained on the development datasets. This observation highlights the robustness of our methods.

We ranked second (out of 2) on all categorization tasks. We ranked third (out of 11) on the event task, and first (out of 3) on the event+ner task. At last, we were the only participant on all knowledge-based tasks.

task	Official results (dev)			
cat	Precision 1.000			
cat+ner	SER	Mism	Ins	Del
	0.702	49.85	127	314
event	Precision	Recall	F-measure	
	0.389	0.644	0.485	
event+ner	SER	P	R	F
	1.486	0.216	0.201	0.208
kb	Mean references 0.7861			
kb+ner	Mean references 0.2074			

Table 5: Official results computed on the development datasets (SER=Slot Error Rate, Mism=Mismatch, Ins=Insertion, Del=Deletion, P=Precision, R=Recall, F=F-measure)

task	Official results (test)			
cat	Precision 0.503			
cat+ner	SER	Mism	Ins	Del
	0.827	198.16	192	455
event	Precision	Recall	F-measure	
	0.388	0.646	0.485	
event+ner	SER	P	R	F
	1.558	0.193	0.192	0.192
kb	Mean references 0.7714			
kb+ner	Mean references 0.2024			

Table 6: Official results computed on the test dataset (SER=Slot Error Rate, Mism=Mismatch, Ins=Insertion, Del=Deletion, P=Precision, R=Recall, F=F-measure)

6 Discussion

6.1 Observations

Additional data A first observation concerns the use of additional data. Increasing the number of annotated files proved to be useful for all machine-learning approaches. In the named entity recognition task—using a CRF system—we gained +3.6 points of F-measure (see table 2). In the relation identification task—using a SVM system—we gained +1.2 points of F-measure for relations based on entities predicted by the CRF, and +0.4 point for relations based on gold standard entities annotations (see table 4). The advantage of using more annotated data is real for all tasks.

Post-processing rules Despite the use of both additional data and over-training reduction function, the CRF model achieved moderate results (F=0.597, see table 2). The use of post-processing rules to refine the CRF outputs slightly increased the overall results (+0.5 points, F=0.602) and mainly impacted the bacteria category (+1.3 points). At last, processing embedded habitat entities with rules improved the overall results (+2.6 points, F=0.628). Using a few post-processing rules increased by +3.1 points the overall results achieved through the CRF model.

Named entity recognition Our strategy based on four steps (additional annotated data, over-training reduction function, post-processing rules, and embedded entities processing) allows us to achieve quite moderate results (F=0.628, see table 2). We failed to identify correctly entities of habitat (F=0.523) while results are higher for both bacteria (F=0.785) and geographical (F=0.739).

Nevertheless, when annotating additional data, we experienced harder work for habitat than for bacteria or geographical. As a consequence, this type of entities is complex for both human annotators and automatic systems.

Categorization The rule-based approach we designed to categorize entities in both the Onto-Biotope ontology and the NCBI taxonomy is quite simple. Since our named entity recognition system obtained moderate results (overall F-measure of 0.628, see table 2), on the categorization task, we achieved better results on the gold standard annotations of entities (overall F-measure of 0.446) than on predictions of entities made by our CRF system (overall F-measure of 0.338, see table 3).

Since we failed to categorize more habitat than bacteria, using default categorization values (see section 4.3) led us to obtain lower precision values for habitat, on both cat+ner (P_{hab}=0.482 vs. P_{bact}=0.714) and cat (P_{hab}=0.518 vs. P_{bact}=0.983) tasks. Moreover, the lowest recall values are also obtained on the categorization of habitat.

6.2 Error analysis

We give in figure 5 a sample of annotations performed by our system on the development dataset (event+ner task).

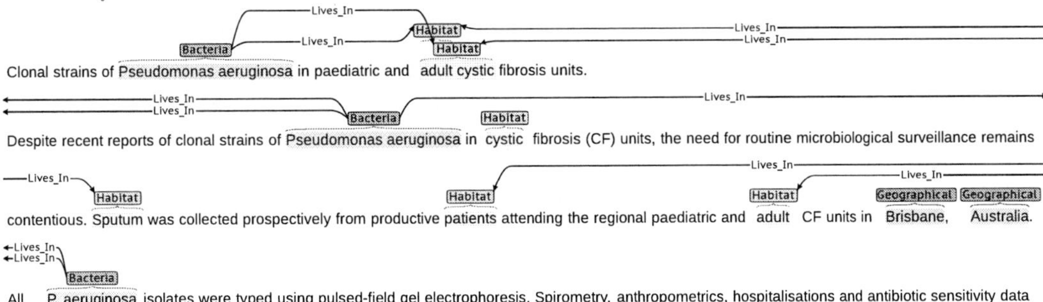

Figure 5: Sample of entities and relations predicted by our system on the development dataset (event+ner task). The first line is the title of the scientific text while other lines are part of the first paragraph

On the NER tasks, our system failed to identify acronyms (*HMDM, HMDMs, PMN, PMNs*), and all discontinuous entities since we chose to not process this kind of entity. False negatives mainly concern habitats: (i) single entities (*paediatric*), (ii) discontinuous entities (*paediatric ... cystic fibrosis units*, *regional ... adult CF units*, and *regional paediatric ... CF units*), and (iii) frontiers errors for which annotations depend on the context (*adult cystic fibrosis units* vs. only *adult cystic* in our sample, *cystic fibrosis (CF) units* vs. *cystic*, or *productive patients* vs. *patients*).

On the categorization tasks, the main errors concern all entities we failed to categorize and for which we gave a default value. Those entities refer to adjectives the system did not process (*pulmonary, duodenal,* etc.) and complex entities (*vacuum- and modified-atmosphere-packed cold-smoked salmon stored at 5 degrees C*, categorized as "vacuum-packed meat" in the reference). As a consequence, each category used as a default value obtained bad results on the development dataset: the NCBI taxonomy entry #210 achieved 34 true positives and 69 false positives while the OntoBiotope entry #002216 achieved 14 true positives, 207 false positives and 11 false negatives.

At last, on the event identification tasks, since there is only one type of relation to identify, the errors concern missing relations and too much relations. Missing relations concern geographical entities (cf. missing relations between *P. aeruginosa* and geographical entities *Brisbane* and *Australia* on figure 5): due to a low number of entities in this category (see table 1), our SVM model failed to learn relations with geographical entities. False positives concern cases where the context between entities prohibits relations (*Neutrophils are resistant to Yersinia*), and annotations done on several lines, including between the content of the title and the content of the other paragraphs (cf. relations between *Pseudomonas aeruginosa* from the first paragraph and habitats *adult cystic* and *adult* from the title).

7   Conclusion

In this paper, we presented the experiments we made while participating in the Bacteria/Biotope track from the 2016 BioNLP Shared-Task. We combined CRF and post-processing rules to identify entities (bacteria, habitat, geographical), including embedded entities, and we used rules based on exact and partial match to normalize the entities in the NCBI taxonomy (bacteria) and the OntoBiotope ontology (habitat). For relation extraction, we used a SVM system based on a basic set of features.

As future work, we plan to deal with discontinuous entities. To process this issue, we consider that a CRF model making the distinction between the pivot and tokens specific to each entity would be useful. As an example, in *throat and nasopharyngeal cultures*, the pivot is *cultures* while specific tokens are *throat* and *nasopharyngeal*. Post-processing rules would bring together tokens so as to produce the final entities (*throat cultures* and *nasopharyngeal cultures*). Our categorization approach to search for partial matches is relatively simple. Future work is needed to provide a better processing of the OntoBiotope ontology, namely, in order to take into account the "is a" relations.

At last, we estimate that using unsupervised learning of relations may provide interesting results, especially to improve the features set used in the SVM model.

References

Sondes Bannour, Laurent Audibert, and Henry Soldano. 2013. Ontology-based semantic annotation: an automatic hybrid rule-based method. In *BioNLP-ST Work Proc*, pages 139–43, Sofia, Bulgaria.

Jari Björne and Tapio Salakoski. 2013. TEES 2.1: Automated Annotation Scheme Learning in the BioNLP 2013 Shared Task. In *BioNLP-ST Work Proc*, pages 16–25, Sofia, Bulgaria.

Robert Bossy, Wiktoria Golik, Zorana Ratkovic, Dialekti Valsamou, Philippe Bessières, and Claire Nédellec. 2015. Overview of the gene regulation network and the bacteria biotope tasks in BioNLP'13 shared task. *BMC Bioinformatics*, 16(Suppl 10):S1.

Robert Bossy, Claire Nédellec, Julien Jourde, and Mouhammadou Ba, 2016. *Guidelines for Annotation of Bacteria Biotopes*. INRA.

Peter F Brown, Vincent J Della Pietra, Peter V de Souza, Jenifer C Lai, and Robert L Mercer. 1992. Class-Based n-gram Models of Natural Language. *Computational Linguistics*, 18(4):467–79.

Vincent Claveau. 2013. IRISA participation to BioNLP-ST 2013: lazy-learning and information retrieval for information extraction tasks. In *BioNLP-ST Work Proc*, pages 188–96, Sofia, Bulgaria.

Louise Deléger, Anne-Laure Ligozat, Cyril Grouin, Pierre Zweigenbaum, and Aurélie Névéol. 2014. Annotation of specialized corpora using a comprehensive entity and relation scheme. In *Proc of LREC*, pages 1267–74, Reykjavik, Iceland.

Louise Deléger, Robert Bossy, Estelle Chaix, Mouhamadou Ba, Arnaud Ferré, Philippe Bessières, and Claire Nédellec. 2016. Overview of the Bacteria Biotope Task at BioNLP Shared Task 2016. In *Proceedings of the 4th BioNLP Shared Task workshop*, Berlin, Germany, August. Association for Computational Linguistics.

Scott Federhen. 2002. The Taxonomy Project. In Johanna McEntyre and Jim Ostell, editors, *The NCBI Handbook*, chapter 4. National Center for Biotechnology Information, Bethesda, MD, 2nd edition.

Cyril Grouin. 2013. Building A Contrasting Taxa Extractor for Relation Identification from Assertions: BIOlogical Taxonomy & Ontology Phrase Extraction System. In *BioNLP-ST Workshop Proc*, pages 144–52, Sofia, Bulgaria.

Thorsten Joachims. 1999. Making large-Scale SVM Learning Practical. In B. Schölkopf, C. Burges, and A. Smola, editors, *Advances in Kernel Methods - Support Vector Learning*, chapter 11, pages 169–184. MIT Press, Cambridge, MA.

İlknur Karadeniz and Arzucan Özgür. 2013. Bacteria Biotope Detection, Ontology-based Normalization, and Relation Extraction using Syntactic Rules. In *BioNLP-ST Work Proc*, pages 170–7, Sofia, Bulgaria.

John D. Lafferty, Andrew McCallum, and Fernando C. N. Pereira. 2001. Conditional Random Fields: Probabilistic models for segmenting and labeling sequence data. In *Proc of ICML*, pages 282–9, Williamstown, MA.

Thomas Lavergne, Olivier Cappé, and François Yvon. 2010. Practical Very Large Scale CRFs. In *Proc of ACL*, pages 504–13, Uppsala, Sweden.

Thomas Lavergne, Cyril Grouin, and Pierre Zweigenbaum. 2015. The contribution of co-reference resolution to supervised relation detection between bacteria and biotopes entities. *BMC Bioinformatics*, 16(Suppl 10):S6.

Percy Liang. 2005. Semi-supervised learning for natural language. Master's thesis, Massachusetts Institute of Technology.

Claire Nédellec. 2016. The OntoBiotope Ontology. Institut National de la Recherche Agronomique.

Pontus Stenetorp, Sampo Pyysalo, Goran Topić, Tomoko Ohta, Sophia Ananiadou, and Juni'chi Tsujii. 2012. BRAT: a Web-based Tool for NLP-Assisted Text Annotation. In *Proc of EACL Demonstrations*, pages 102–7, Avignon, France. ACL.

Vladimir N Vapnik. 1995. *The Nature of Statistical Learning Theory*. Springer-Verlag.

Karin Verspoor, Antonio Jimeno Yepes, Lawrence Cavedon, Tara McIntosh, Asha Herten-Crabb, Zo Thomas, and John-Paul Plazzer. 2013. Annotating the Biomedical Literature for the Human Variome. In *Database: The Journal of Biological Databases and Curation*.

Deep Learning with Minimal Training Data: TurkuNLP Entry in the BioNLP Shared Task 2016

Farrokh Mehryary[1,3], **Jari Björne**[2,3], **Sampo Pyysalo**[4], **Tapio Salakoski**[2,3] and **Filip Ginter**[2,3]

[1]University of Turku Graduate School (UTUGS)
[2]Turku Centre for Computer Science (TUCS)
[3]Department of Information Technology, University of Turku
Faculty of Mathematics and Natural Sciences, FI-20014, Turku, Finland
[4]Language Technology Lab, DTAL, University of Cambridge
`firstname.lastname@utu.fi, sampo@pyysalo.net`

Abstract

We present the TurkuNLP entry to the BioNLP Shared Task 2016 Bacteria Biotopes event extraction (BB3-event) subtask. We propose a deep learning-based approach to event extraction using a combination of several Long Short-Term Memory (LSTM) networks over syntactic dependency graphs. Features for the proposed neural network are generated based on the shortest path connecting the two candidate entities in the dependency graph. We further detail how this network can be efficiently trained to have good generalization performance even when only a very limited number of training examples are available and part-of-speech (POS) and dependency type feature representations must be learned from scratch. Our method ranked second among the entries to the shared task, achieving an F-score of 52.1% with 62.3% precision and 44.8% recall.

1 Introduction

The BioNLP Shared Task 2016 was the fourth in the series to focus on event extraction, an information extraction task targeting structured associations of biomedical entities (Kim et al., 2009; Ananiadou et al., 2010). The 2016 task was also the third to include a Bacteria Biotopes (BB) subtask focusing on microorganisms and their habitats (Bossy et al., 2011). Here, we present the TurkuNLP entry to the BioNLP Shared Task 2016 Bacteria Biotope event extraction (BB3-event) subtask. Our approach builds on proven tools and ideas from previous tasks and is novel in its application of deep learning methods to biomedical event extraction.

The BB task was first organized in 2011, then consisting of named entity recognition (NER) targeting mentions of bacteria and locations, followed by the detection of two types of relations involving these entities (Bossy et al., 2011). Three teams participated in this task, with the best F-score of 45% achieved by the INRA Bibliome group with the Alvis system, which used dictionary mapping, ontology inference and semantic analysis for NER, and co-occurrence-based rules for detecting relations between the entities (Ratkovic et al., 2011). The 2013 BB task defined three subtasks (Nédellec et al., 2013), the first one concerning NER, targeting bacteria habitat entities and their normalization, and the other two subtasks involving relation extraction, the task targeted also by the system presented here. Similarly to the current BB3-event subtask, the 2013 subtask 2 concerned only relation extraction, and subtask 3 extended this with NER. Four teams participated in these tasks, with the UTurku TEES system achieving the first places with F-scores of 42% and 14% (Björne and Salakoski, 2013).

We next present the 2016 BB3-event subtask and its data and then proceed to detail our method, its results and analysis. We conclude with a discussion of considered alternative approaches and future work.

2 Task and Data

In this section, we briefly present the BB3-event task and the statistics of the data that has been used for method development and optimization, as well as for test set prediction.

Although the BioNLP Shared Task has introduced an event representation that can capture associations of arbitrary numbers of participants in complex, recursive relationships, the BB3-event task follows previous BB series subtasks in ex-

Proceedings of the 4th BioNLP Shared Task Workshop, pages 73–81,
Berlin, Germany, August 13, 2016. ©2016 Association for Computational Linguistics

	Train	Devel	Test
Total sentences	527	319	508
Sentences w/examples	158	117	158
Sentences w/o examples	369	202	350
Total examples	524	506	534
Positive examples	251	177	-
Negative examples	273	329	-

Table 1: BB3-event data statistics. (The relation annotations of the test set have not been released.)

clusively marking directed binary associations of exactly two entities. For the purposes of machine learning, we thus cast the BB3-event task as binary classification taking either a (BACTERIA, HABITAT) or a (BACTERIA, GEOGRAPHICAL) entity pair as input and predicting whether or not a *Lives-in* relation holds between the BACTERIA and the location (HABITAT or GEOGRAPHICAL).

Our approach builds on the shortest dependency path between each pair of entities. However, while dependency parse graphs connect words to others in the same sentence, a number of annotated relations in the data involve entities appearing in different sentences, where no connecting path exists. Such cross-sentence associations are known to represent particular challenges for event extraction systems, which rarely attempt their extraction (Kim et al., 2011). In this work, we simply exclude cross-sentence examples from the data. This elimination procedure resulted in the removal of 106 annotated relations from the training set and 62 annotated relations from the development set.

The examples that we use for the training, optimization and development evaluation of our method are thus a subset of those in the original data.[1] When discussing the training, development and test data, we refer to these filtered sets throughout this manuscript. The statistics of the task data after this elimination procedure are summarized in Table 1. Note that since there are various ways of converting the shared task annotations into examples for classification, the numbers we report here may differ from those reported by other participating teams.

---

[1] Official evaluation results on the test data are of course comparable to those of other systems: any cross-sentence relations in the test data count against our submission as false negatives.

3 Method

We next present our method in detail. Preprocessing is first discussed in Section 3.1. Section 3.2 then explains how the shortest dependency path is used, and the architecture of the proposed deep neural network is presented in Section 3.3. Section 3.4 defines the classification features and embeddings for this network. Finally, in Section 3.5 we discuss the training and regularization of the network.

3.1 Preprocessing

We use the TEES system, previously developed by members of the TurkuNLP group (Björne and Salakoski, 2013), to run a basic preprocessing pipeline of tokenization, POS tagging, and parsing, as well as to remove cross-sentence relations. Like our approach, TEES targets the extraction of associations between entities that occur in the same sentence. To support this functionality, it can detect and eliminate relations that cross sentence boundaries in its input. We use this feature of TEES as an initial preprocessing step to remove such relations from the data.

To obtain tokens, POS tags and parse graphs, TEES uses the BLLIP parser (Charniak and Johnson, 2005) with the biomedical domain model created by McClosky (2010). The phrase structure trees produced by the parser are further processed with the Stanford conversion tool (de Marneffe et al., 2006) to create dependency graphs. The Stanford system can produce several variants of the Stanford Dependencies (SD) representation. Here, we use the *collapsed* variant, which is designed to be useful for information extraction and language understanding tasks (de Marneffe and Manning, 2008).

3.2 Shortest Dependency Path

The syntactic structure connecting two entities e_1 and e_2 in various forms of syntactic analysis is known to contain most of the words relevant to characterizing the relationship $R(e_1, e_2)$, while excluding less relevant and uninformative words.

This observation has served as the basis for many successful relation extraction approaches in both general and biomedical domain NLP (Bunescu and Mooney, 2005; Airola et al., 2008; Nguyen et al., 2009; Chowdhury et al., 2011). The TEES system also heavily relies on the shortest dependency path for defining and extracting features (Björne et al., 2012; Björne and Salakoski, 2013). Recently, this idea was applied in an LSTM-based relation extraction system by Xu et al. (2015). Since the dependency parse is directed (i.e. the path from e_1 to e_2 differs from that from e_2 to e_1), they separate the shortest dependency path into two sub-paths, each from an entity to the common ancestor of the two entities, generate features along the two sub-paths, and feed them into different LSTM networks, to process the information in a direction sensitive manner.

To avoid doubling the number of LSTM chains (and hence the number of weights), we convert the dependency parse to an undirected graph, find the shortest path between the two entities (BACTERIA and HABITAT/GEOGRAPHICAL), and always proceed from the BACTERIA entity to the HABITAT/GEOGRAPHICAL entity when generating features along the shortest path, regardless of the order of the entity mentions in the sentence. With this approach, there is a single LSTM chain (and set of LSTM weights) for every feature set, which is more effective when the number of training examples is limited.

There is a subtle and important point to be addressed here: as individual entity mentions can consist of several (potentially discontinuous) tokens, the method must be able to select which word (i.e. single token) serves as the starting/ending point for paths through the dependency graph. For example, in the following training set sentence, *"biotic surfaces"* is annotated as a HABITAT entity:

> *"We concluded that S. marcescens MG1 utilizes different regulatory systems and adhesins in attachment to **biotic** and abiotic **surfaces** [...]"*

As this mention consists of two (discontinuous) tokens, it is necessary to decide whether the paths connecting this entity to BACTERIA mentions (e.g., *"S. marcescens MG1"*) should end at *"biotic"* or *"surfaces"*. This problem has fortunately been addressed in detail in previous work, allowing us to adopt the proven solution proposed by Björne et al. (2012) and implemented in the TEES system, which selects the *syntactic head*, i.e. the root token of the dependency parse sub-tree covering the entity, for any given multi-token entity. Hence, in the example above, the token *"surfaces"* is selected and used for finding the shortest dependency paths.

3.3 Neural Network Architecture

While recurrent neural networks (RNNs) are inherently suitable for modeling sequential data, standard RNNs suffer from the *vanishing or exploding gradients* problem: if the network is deep, during the back-propagation phase the gradients may either decay exponentially, causing learning to become very slow or stop altogether (*vanishing* gradients); or become excessively large, causing the learning to diverge (*exploding* gradients) (Bengio et al., 1994). To avoid this issue, we make use of Long Short-Term Memory (LSTM) units, which were proposed to address this problem (Hochreiter and Schmidhuber, 1997).

We propose an architecture centered around three RNNs (chains of LSTM units): one representing *words*, the second *POS tags*, and the third *dependency types* (Figure 1). For a given example, the sequences of words, POS tags and dependency types on the shortest dependency path from the BACTERIA mention to the HABITAT/GEOGRAPHICAL mention are first mapped into vector sequences by three separate embedding lookup layers. These word, POS tag and dependency type vector sequences are then input into the three RNNs. The outputs of the last LSTM unit of each of the three chains are then concatenated and the resulting higher-dimensional vector input to a fully connected hidden layer. The hidden layer finally connects to a single-node binary classification layer.

Based on experiments on the development set, we have set the dimensionality of all LSTM units and the hidden layer to 128. The sigmoid activation function is applied on the output of all LSTM units, the hidden layer and the output layer.

3.4 Features and Embeddings

We next present the different embeddings defining the primary features of our model. In addition to the embeddings, we use a binary feature which has the value 0 if the corresponding location is a GEOGRAPHICAL entity and 1 if it is a HABITAT entity. This input is directly concatenated with the LSTM outputs and fed into the hidden layer. We noticed this signal slightly improves classification performance, resulting in a less than 1 percentage point increase of the F-score.

3.4.1 Word embeddings

We initialize our word embeddings with vectors induced using six billion words of biomedical

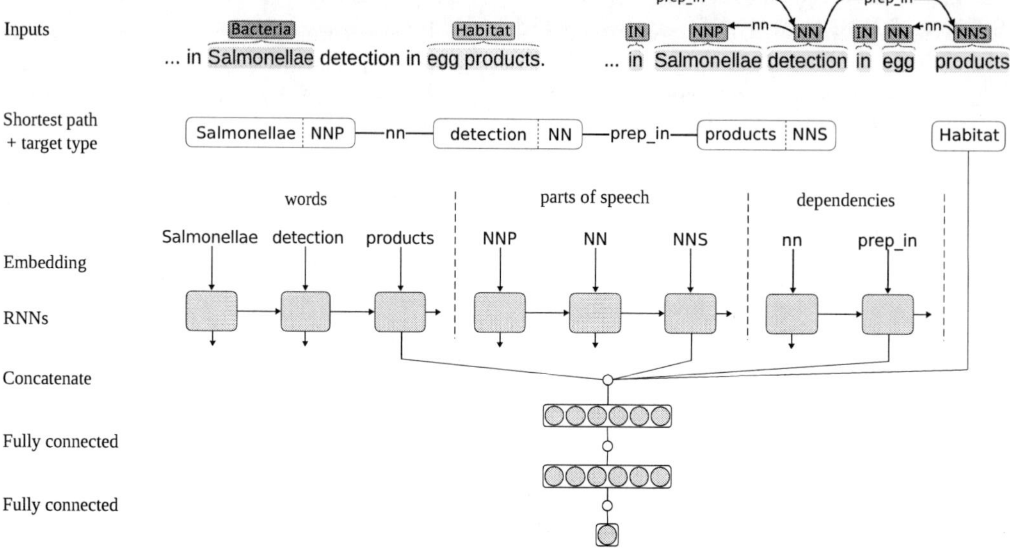

Figure 1: Proposed network architecture.

scientific text, namely the combined texts of all PubMed titles and abstracts and PubMed Central Open Access (PMC OA) full text articles available as of the end of September 2013.[2] These 200-dimensional vectors were created by Pyysalo et al. (2013) using the `word2vec` implementation of the *skip-gram* model (Mikolov et al., 2013).

To reduce the memory requirements of our method, we only use the vectors of the 100,000 most frequent words to construct the embedding matrix. Words not included in this vocabulary are by default mapped to a shared, randomly initialized unknown word vector. As an exception, out of vocabulary BACTERIA mentions are instead mapped to the vector of the word *"bacteria"*. Based on development set experiments we estimate that this special-case processing improved the F-score by approximately 1% point.

3.4.2 POS embeddings

Our POS embedding matrix consists of a 100-dimensional vector for each of the POS tags in the Penn Treebank scheme used by the applied tagger. We do not use pre-trained POS vectors but instead initialize the embeddings randomly at the beginning of the training phase.

[2]Available from `http://bio.nlplab.org/`

3.4.3 Dependency type embeddings

Typed dependencies – the edges of the parse graph – represent directed grammatical relations between the words of a sentence. The sequence of dependencies on the shortest path between two entities thus conveys highly valuable information about the nature of their relation.

We map each dependency type in the collapsed SD representation into a randomly initialized 350-dimensional vector (size set experimentally). Note that in the applied SD variant, prepositions and conjunctions become part of collapsed dependency types (de Marneffe et al., 2006), as illustrated in Figure 2.

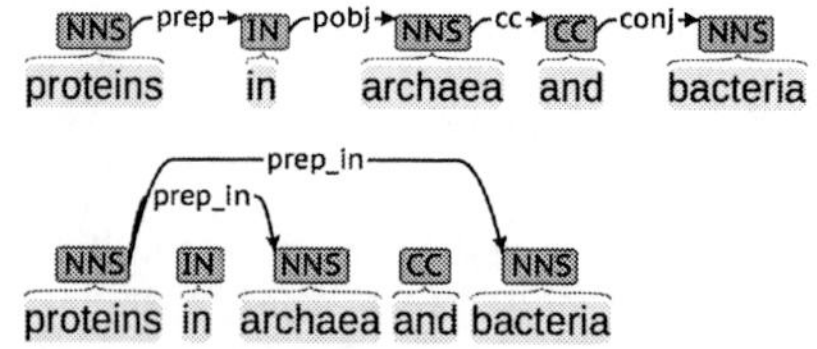

Figure 2: Basic (top) and collapsed (bottom) Stanford Dependency representations

As the collapsed dependencies thus incorporate preposition and conjunction words into the grammatical relations themselves, the set of dependency types is somewhat open-ended. To account for this, all preposition/conjunction dependency types not observed in the training and development sets are mapped to the vectors for the general preposition and conjunction types `prep` and `conj`, respectively.

3.5 Training and Regularization

We use *binary cross-entropy* as the objective function and the *Adam* optimization algorithm with the parameters suggested by Kingma and Ba (2014) for training the network. We found that this algorithm yields considerably better results than the conventional stochastic gradient descent in terms of classification performance.

During training, the randomly initialized POS and dependency type embeddings are trained and the pre-trained word embeddings fine-tuned by back-propagation using the supervised signal from the classification task at hand.

Determining how long to train a neural network model for is critically important for its generalization performance. If the network is *under-trained*, model parameters will not have converged to good values. Conversely, *over-training* leads to overfitting on the training set. A conventional solution is early stopping, where performance is evaluated on the development set after each set period of training (e.g. one pass through the training set, or *epoch*) to decide whether to continue or stop the training process. A simple rule is to continue while the performance on the development set is improving. By repeating this approach for 15 different runs with different initial random initializations of the model, we experimentally concluded that the optimal length of training is four epochs. Overfitting is a serious problem in deep neural networks with a large number of parameters. To reduce overfitting, we experimented with several regularization methods including the l_1 weight regularization penalty (*LASSO*) and the l_2 weight decay (*ridge*) penalty on the hidden layer weights. We also tried the dropout method (Srivastava et al., 2014) on the output of LSTM chains as well as on the output of the hidden layer, with a dropout rate of 0.5. Out of the different combinations, we found the best results when applying dropout after the hidden layer. This is the only regularization method used in the final method.

4 Results

4.1 Overcoming Variance

At the beginning of training, the weights of the neural network are initialized randomly. As we are

Run	Recall	Precision	F-score
12	76.3	60.3	**67.3**
14	71.2	63.0	66.8
13	75.7	59.3	66.5
10	78.0	56.3	65.4
3	**80.8**	54.0	64.7
15	79.1	54.3	64.4
1	66.1	62.2	64.1
11	65.0	62.8	63.9
2	67.8	59.4	63.3
5	55.9	69.7	62.1
7	57.6	66.7	61.8
9	53.1	70.2	60.5
8	50.9	74.4	60.4
6	50.3	73.6	59.7
4	46.9	**78.3**	58.7
$\bar{x}$	65.0	64.3	63.3
σ	11.3	7.3	2.6

Table 2: Development set results for 15 repetitions with different initial random initializations with mean ($\bar{x}$) and standard deviation (σ). Results are sorted by F-score.

only using pre-trained embeddings for words, this random initialization applies also to the POS and dependency type embeddings. Since the number of weights is high and the training set is very small (only 524 examples), the initial random state of the model can have a significant impact on the final model and its generalization performance. Limited numbers of training examples are known to represent significant challenges for leveraging the full power of deep neural networks, and we found this to be the case also in this task.

To study the influence of random effects on our model, we evaluate it with 15 different random initializations, training each model for four epochs on the training data and evaluating on the development set using the standard precision, recall and F-score metrics. Table 2 shows the obtained results. We find that the primary evaluation metric, the F-score, varies considerably, ranging from 58.7% to 67.3%. This clearly illustrates the extent to which the random initialization can impact the performance of the model on unseen data. While the method is shown to obtain on average an F-score of 63.3% on the development set, it must be kept in mind that given the standard deviation of 2.6, individual trained models may perform substantially better (or worse). It is also important to note that due to the small size of the development

Threshold (t)	Recall	Precision	F-score
1	**83.6**	53.2	65.1
2	79.7	54.0	64.4
3	78.5	57.0	66.0
4	78.0	59.0	**67.2**
5	75.7	60.1	67.0
6	70.6	60.7	65.3
7	67.8	61.5	64.5
8	65.5	62.0	63.7
9	62.2	65.5	63.8
10	58.2	66.5	62.1
11	57.1	69.7	62.7
12	52.5	70.5	60.2
13	51.4	72.8	60.3
14	48.6	74.8	58.9
15	45.2	**80.0**	57.8

Table 3: Development set results for voting based on the predictions of the 15 different classifiers. Best results for each metric shown in bold.

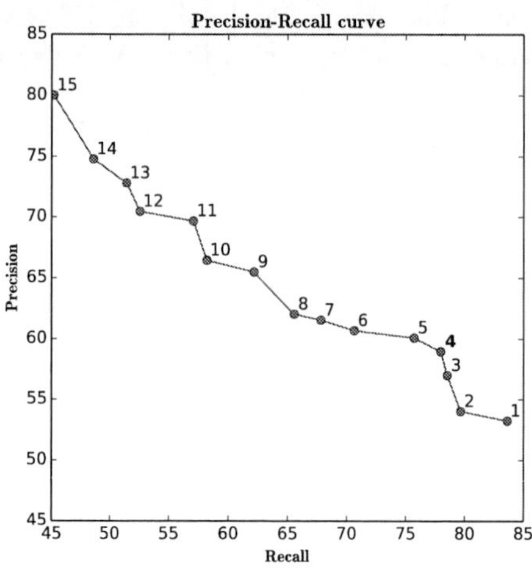

Figure 3: Precision-recall curve for different values of the threshold t (shown as labels on the curve).

set, individual models that achieved high performance in this experiment will not necessarily generalize well to unseen data.

To deal with these issues, we introduce a straightforward voting procedure that aggregates the prediction outputs of the 15 classifiers based on a given threshold value $t \in \{1, \ldots, 15\}$:

1. For each example, predict outputs with the 15 models;

2. If at least t outputs are positive, label the example positive, otherwise label it negative.

Clearly, the most conservative threshold is $t = 15$, where a relation is voted to exist only if *all* the 15 classifiers have predicted it. Conversely, the least conservative threshold is $t = 1$, where a relation is voted to hold if *any* classifier has predicted it.

The development set results for the voting algorithm with different threshold values are given in Table 3. As expected, the threshold $t = 1$ produces the highest recall (83.6%) with the lowest precision (53.2%). With increasing values of t, precision increases while recall drops, and the highest precision (80.0%) is achieved together the lowest recall of (45.2%) with $t = 15$. The best F-score is obtained with $t = 4$, where an example is labeled positive if at least four classifiers have predicted it to be positive and negative otherwise. Figure 3 shows the precision-recall curve for these 15 threshold values.

As is evident from these results, the voting algorithm can be used for different purposes. If the aim is to obtain the best overall performance, we can investigate which threshold produces the highest F-score (here $t = 4$) and select that value when making predictions for unseen data (e.g., the test set). Alternatively, for applications that specifically require high recall or high precision, a different threshold value can be selected to optimize the desired metric.

To assess the performance of the method on the full, *unfiltered* development set that includes also cross-sentence relations, we selected the threshold value $t = 4$ and submitted the aggregated prediction results to the official Shared Task evaluation server. The method achieved an F-score of 60.0% (60.9% precision and 59.3% recall), 7.2% points below the result for our internal evaluation with filtered data (Table 3).

4.2 Test Set Evaluation

For evaluation of the test set, we applied the proposed model with the voting approach presented above): 15 neural network models with different random initializations were trained for 4 epochs on the combination of the training and the development sets. Each trained model was then used to produce one set of predictions for the test set. To obtain the final test set predictions, the outputs of the 15 classifiers were aggregated using the voting algorithm with a threshold $t = 4$.

Our method achieved an F-score of 52.1% with a recall of 44.8% and precision of 62.3%, ranking second among the entries to the shared task. We again emphasize that our approach ignored all potential relations between entities belonging to different sentences, which may in part explain the comparatively low recall.

4.3 Runtime Performance and Technical Details

We implemented the model using the Python programming language (v2.7) with Keras, a model-level deep learning library (Chollet, 2015). All network parameters not explicitly discussed above were left to their defaults in Keras. The Theano tensor manipulation library (Bastien et al., 2012) was used as the backend engine for Keras. Computations were run on a single server computer equipped with a GPU.[3] All basic python processing, including e.g. file manipulation, the TEES pipeline and our voting algorithm, was run on a single CPU core, while all neural network related calculations (training, optimization, predictions) were run on the GPU, using the CUDA toolkit version 5.0.

The training process takes about 10 minutes, including model building and 4 epochs of training the network on the training set, but excluding preprocessing and the creation and loading of the input network. Prediction of the development set using a trained model with fully prepared inputs is very fast, taking only about 10 seconds. Finally, the voting algorithm executes in less than a minute for all 15 thresholds.

We note that even though the proposed approach involving 15 rounds of training, prediction and result aggregation might seem to be impractical for large-scale real-word applications (e.g., extracting bacteria-location relations from all PubMed abstracts), it is quite feasible in practice, as the time-consuming training process only needs to be done once, and prediction on unseen data is quite fast.

4.4 Other Architectures

In this section, we discuss alternative approaches that we considered and the reasons why they were rejected in favor of that described above.

One popular and proven method for relation extraction is to use three groups of features, based on the observation that the words preceding the first entity, the words between the entities, and those after the second entity serve different roles in deciding whether or not the entities are related (Bunescu and Mooney, 2006). Given a sentence $S = w_1, ..., e_1, ..., w_i, ..., e_2, ..., w_n$ with entities e_1 and e_2, one can represent the sentence with three groups of words: $\{before\}e_1\{middle\}e_2\{after\}$ (e_1 and e_2 can also be included in the groups). The similarity of two examples represented in this way can be compared using e.g. *sub-sequence kernels* at word level (Bach and Badaskar, 2007). Bunescu and Mooney (2006) utilize three subkernels matching combinations of the before, middle and after sequences of words, with a combined kernel that is simply the sum of the subkernels. This kernel is then used with support vector machines for relation extraction. Besides the words, other features such as the corresponding POS tags and entity types can also be incorporated into such kernel functions to further improve the representation.

We adapted this idea to deep neural networks. We started with the simplest architecture, which contains 3 LSTM networks. Instead of generating features based on the shortest path, each LSTM receives inputs based on the sequence of the words seen in each of the *before*, *middle*, and *after* groups, where the word embeddings are the only features used for classification. Similar to the architecture discussed in Section 3.3, the outputs of the last LSTM units in each chain are concatenated, and the resulting higher-dimensional vector is then fed into a fully connected hidden layer and then to the output layer. This approach has a major advantage over the shortest dependency path, in particular for large-scale applications: parsing, *the most time-consuming part* in the relation extraction pipeline, is no longer required.

Unfortunately, our internal evaluation on the development set showed that this model failed to achieve results comparable to those of the shortest dependency path model, only reaching an F-score of about 57%. Hence, we attempted to use more features by adding 3 or 6 additional LSTM chains to the model, for POS or/and dependency type embeddings. Even in these cases, the F-scores only varied in the range of 57% to about 63% (for different random initializations). We conclude

---

[3] In detail: two 6-core Intel® Xeon® *E5-2620* processors, 32 gigabytes of main memory, and one NVIDIA® TESLA™ *C2075* companion processor with 448 CUDA cores and 6 gigabytes of memory.

that even though not requiring parsing is a benefit in these approaches, our experiments suggest that they are not capable of reaching performance comparable to methods that use the syntactic structure of sentences.

5 Conclusions and Future work

We have presented the entry of the TurkuNLP team to the Bacteria Biotope event extraction (BB3-event) sub-task of the BioNLP Shared Task 2016. Our method is based on a combination of LSTM networks over syntactic dependency graphs. The features for the network are derived from the POS tags, dependency types, and word forms occurring on the shortest dependency path connecting the two candidate entities (BACTERIA and HABITAT/GEOGRAPHICAL) in the collapsed Stanford Dependency graph.

We initialize word representations using pre-trained vectors created using six billion words of biomedical text (PubMed and PMC documents). During training, the pre-trained word embeddings are fine-tuned while randomly initialized POS and dependency type representations are trained from scratch. We showed that as the number of training examples is very limited, the random initialization of the network can considerably impact the quality of the learned model. To address this issue, we introduced a voting approach that aggregates the outputs of differently initialized neural network models. Different aggregation thresholds can be used to select different precision-recall trade-offs. Using this method, we showed that our proposed deep neural network can be efficiently trained to have good generalization for unseen data even with minimal training data. Our method ranked second among the entries to the shared task, achieving an F-score of 52.1% with 62.3% precision and 44.8% recall.

There are a number of open questions regarding our model that we hope to address in future work. First, we observed how the initial random state of the model can impact its final performance on unseen data. It is interesting to investigate whether (and to what extent) pre-training the POS and dependency type embeddings can address this issue. One possible approach would be to apply the method to similar biomedical relation extraction tasks that include larger corpora than the BB3-event task (Pyysalo et al., 2008) and use the learned POS and dependency embeddings for initialization for this task. This could also establish to what extent pre-training these representations can boost the F-score.

Second, it will be interesting to study how the method performs with different amounts of training data. On one hand, we can examine to what extent the training corpus size can be reduced without compromising the ability of the proposed network to learn the classification task; on the other, we can explore how this deep learning method compares with previously proposed state-of-the-art biomedical relation extraction methods on larger relation extraction corpora.

Third, the method and task represent an opportunity to study how the word embeddings used for initialization impact relation extraction performance and in this way assess the benefits of different methods for creating word embeddings in an extrinsic task with real-world applications.

Finally, it is interesting to investigate different methods to deal with cross-sentence relations. Here we ignored all potential relations where the entities are mentioned in different sentences as there is no path connecting tokens across sentences in the dependency graph. One simple method that could be considered is to create an artificial "paragraph" node connected to all sentence roots to create such paths (cf. e.g. Melli et al. (2007)).

We aim to address these open questions and further extensions of our model in future work.

Acknowledgments

We would like to thank Kai Hakala, University of Turku, for his technical suggestions for the pipeline development and the anonymous reviewers for their insightful comments and suggestions. We would also like to thank the CSC IT Center for Science Ltd. for computational resources. This work has been supported in part by Medical Research Council grant MR/M013049/1.

References

Antti Airola, Sampo Pyysalo, Jari Björne, Tapio Pahikkala, Filip Ginter, and Tapio Salakoski. 2008. All-paths graph kernel for protein-protein interaction extraction with evaluation of cross-corpus learning. *BMC bioinformatics*, 9(11):1.

Sophia Ananiadou, Sampo Pyysalo, Junichi Tsujii, and Douglas B Kell. 2010. Event extraction for systems biology by text mining the literature. *Trends in biotechnology*, 28(7):381–390.

Nguyen Bach and Sameer Badaskar. 2007. A review of relation extraction. *Language Technologies Institute, Carnegie Mellon University*.

Frédéric Bastien, Pascal Lamblin, Razvan Pascanu, James Bergstra, Ian J. Goodfellow, Arnaud Bergeron, Nicolas Bouchard, and Yoshua Bengio. 2012. Theano: new features and speed improvements. In *Proc. Deep Learning and Unsupervised Feature Learning NIPS 2012 Workshop*.

Yoshua Bengio, Patrice Simard, and Paolo Frasconi. 1994. Learning long-term dependencies with gradient descent is difficult. *IEEE Transactions on Neural Networks*, 5(2):157–166.

Jari Björne and Tapio Salakoski. 2013. TEES 2.1: Automated annotation scheme learning in the bionlp 2013 shared task. In *Proc. BioNLP Shared Task*, pages 16–25.

Jari Björne, Filip Ginter, and Tapio Salakoski. 2012. University of Turku in the BioNLP'11 shared task. *BMC Bioinformatics*, 13(S-11):S4.

Robert Bossy, Julien Jourde, Philippe Bessières, Maarten van de Guchte, and Claire Nédellec. 2011. Bionlp shared task 2011: Bacteria biotope. In *Proc. BioNLP Shared Task*, pages 56–64.

Razvan C. Bunescu and Raymond J. Mooney. 2005. A shortest path dependency kernel for relation extraction. In *Proc. HLT-EMNLP*, pages 724–731.

Razvan Bunescu and Raymond J. Mooney. 2006. Subsequence kernels for relation extraction. In *Proc. NIPS*.

Eugene Charniak and Mark Johnson. 2005. Coarse-to-fine N-best parsing and maxent discriminative reranking. In *Proc. ACL*, pages 173–180.

Franois Chollet. 2015. Keras. `https://github.com/fchollet/keras`.

Faisal Mahbub Chowdhury, Alberto Lavelli, and Alessandro Moschitti. 2011. A study on dependency tree kernels for automatic extraction of protein-protein interaction. In *Proc. BioNLP 2011*, pages 124–133.

Marie-Catherine de Marneffe and Christopher D. Manning. 2008. The stanford typed dependencies representation. In *Proc. CrossParser*, pages 1–8.

Marie-Catherine de Marneffe, Bill MacCartney, and Christopher D. Manning. 2006. Generating typed dependency parses from phrase structure parses. In *Proc. LREC-2006*, pages 449–454.

Sepp Hochreiter and Jürgen Schmidhuber. 1997. Long short-term memory. *Neural Comput.*, 9(8):1735–1780.

Jin-Dong Kim, Tomoko Ohta, Sampo Pyysalo, Yoshinobu Kano, and Jun'ichi Tsujii. 2009. Overview of BioNLP'09 shared task on event extraction. In *Proc. BioNLP Shared Task*, pages 1–9.

Jin-Dong Kim, Tomoko Ohta, Sampo Pyysalo, Yoshinobu Kano, and Junichi Tsujii. 2011. Extracting bio-molecular events from literature - the BioNLP'09 shared task. *Computational Intelligence*, 27(4):513–540.

Diederik P. Kingma and Jimmy Ba. 2014. Adam: A method for stochastic optimization. *CoRR*, abs/1412.6980.

David McClosky. 2010. *Any Domain Parsing: Automatic Domain Adaptation for Natural Language Parsing*. Ph.D. thesis.

Gabor Melli, Martin Ester, and Anoop Sarkar. 2007. Recognition of multi-sentence n-ary subcellular localization mentions in biomedical abstracts. In *Proceedings of LBM*.

Tomas Mikolov, Ilya Sutskever, Kai Chen, Greg S Corrado, and Jeff Dean. 2013. Distributed representations of words and phrases and their compositionality. In *Advances in Neural Information Processing Systems 26*, pages 3111–3119.

Claire Nédellec, Robert Bossy, Jin-Dong Kim, Jung-Jae Kim, Tomoko Ohta, Sampo Pyysalo, and Pierre Zweigenbaum. 2013. Overview of BioNLP shared task 2013. In *Proc. BioNLP Shared Task*, pages 1–7.

Truc-Vien T Nguyen, Alessandro Moschitti, and Giuseppe Riccardi. 2009. Convolution kernels on constituent, dependency and sequential structures for relation extraction. In *Proc. EMNLP*, pages 1378–1387.

Sampo Pyysalo, Antti Airola, Juho Heimonen, Jari Björne, Filip Ginter, and Tapio Salakoski. 2008. Comparative analysis of five protein-protein interaction corpora. *BMC bioinformatics*, 9(3):1.

Sampo Pyysalo, Filip Ginter, Hans Moen, Tapio Salakoski, and Sophia Ananiadou. 2013. Distributional semantic resources for biomedical text mining. In *Proc. LBM*, pages 39–44.

Zorana Ratkovic, Wiktoria Golik, Pierre Warnier, Philippe Veber, and Claire Nédellec. 2011. BioNLP 2011 task Bacteria Biotope: The Alvis system. In *Proc. BioNLP Shared Task*, pages 102–111.

Nitish Srivastava, Geoffrey Hinton, Alex Krizhevsky, Ilya Sutskever, and Ruslan Salakhutdinov. 2014. Dropout: A simple way to prevent neural networks from overfitting. *Journal of Machine Learning Research*, 15:1929–1958.

Yan Xu, Lili Mou, Ge Li, Yunchuan Chen, Hao Peng, and Zhi Jin. 2015. Classifying relations via long short term memory networks along shortest dependency paths. In *Proc. EMNLP*, pages 1785–1794.

SeeDev Binary Event Extraction using SVMs and a Rich Feature Set

Nagesh C. Panyam, Gitansh Khirbat,
Karin Verspoor, Trevor Cohn and Kotagiri Ramamohanarao
Department of Computing and Information Systems,
The University of Melbourne,
Australia
{npanyam, gkhirbat}@student.unimelb.edu.au
{karin.verspoor, t.cohn, kotagiri}@unimelb.edu.au

Abstract

This paper describes the system details and results of the participation of the team from the University of Melbourne in the SeeDev binary event extraction of BioNLP-Shared Task 2016. This task addresses the extraction of genetic and molecular mechanisms that regulate plant seed development from the natural language text of the published literature. In our submission, we developed a system[1] using a support vector machine classifier with linear kernel powered by a rich set of features. Our system achieved an F1-score of 36.4%.

1 Introduction

One of the biggest research challenges faced by the agricultural industry is to understand the molecular network underlying the regulation of seed development. Different tissues involving complex genetics and various environmental factors are responsible for the healthy development of a seed. A large body of research literature is available containing this knowledge. The SeeDev binary relation extraction subtask of the BioNLP Shared Task 2016 (Chaix et al., 2016) focuses on extracting relations or events that involve two biological entities as expressed in full-text publication articles. The task represents an important contribution to the broader problem of biomedical relation extraction.

Similar to previous BioNLP shared tasks in 2009 and 2011 (Kim et al., 2009; Kim et al., 2011), this task focuses on molecular information extraction. The task organisers provided paragraphs from manually selected full text publications on seed development of *Arabidopsis thaliana*

annotated with mentions of biological entities like *proteins* and *genes*, and binary relations like *Exists_In_Genotype* and *Occurs_In_Genotype*. The participants are asked to extract binary relations between entities in a given paragraph.

Several approaches have been proposed to extract biological events from text (Ohta et al., 2011; Liu et al., 2013). Broadly, these approaches can be categorized into two main groups, namely rule-based and machine learning (ML) based approaches. Rule-based approaches consist of a set of rules that are manually defined or semi-automatically inferred from the training data (Abacha and Zweigenbaum, 2011). To extract events from text, first event triggers are detected using a dictionary, then the defined rules are applied over rich representations such as dependency parse trees, to extract the event arguments. On the other hand, ML-based approaches (Miwa et al., 2010) are characterized by learning algorithms such as classification to extract event arguments. Further, they employ various features computed from the textual or syntactic properties of the input text.

This article explains our SeeDev binary relation extraction system in detail. We describe the rich feature set and classifier setup employed by our system that helped achieve the second best F1-score of 36.4% in the shared task.

2 Approach

The seedev task involves extraction of 22 different binary events over 16 entity types. Entity mentions within a sentence and the events between them are provided in the gold standard annotations. In the rest of the article, we refer to an event with two entity arguments as simply a binary relation.

We treat relation identification as a supervised classification problem and created 22 separate

---

[1] Source: https://github.com/unimelbbionlp/BioNLPST2016/

Proceedings of the 4th BioNLP Shared Task Workshop, pages 82–87,
Berlin, Germany, August 13, 2016. ©2016 Association for Computational Linguistics

classifiers denoted as $C_1, C_2, \ldots, C_{22}$, specific to relations $r_1, r_2, \ldots, r_{22}$, respectively. This design choice was motivated by two important aspects, namely vocabulary and relation type signature. We describe them below:

Vocabulary According to the annotation guidelines document, it is clear that different relations are expressed using different vocabulary. For example, "encode" is in the vocabulary of "Transcribes_Or_Translates_To" and "phosphorylate" is in the vocabulary of "Regulation_Of_Molecule_Activity". We hypothesize that treating the vocabulary as a set of trigger words for its corresponding relation would be beneficial. Therefore, we built 22 separate classifiers for each relation type, with vocabulary as a relation specific feature. Given an entity pair (e_a, e_b), we test it with the classifier C_i to detect if the relation r_i holds between e_a and e_b.

Relation type signature Relations are associated with entity type argument signatures, which specify the list of allowed entity types for each argument position. For example, the event "Protein Complex Composition" requires the first entity argument to be one of these four entity types {"Protein", "Protein Family", "Protein Complex", "Protein Domain"} and the second argument to be "Protein Complex". Alternately, relation argument signatures can be used as a filter that specifies the list of invalid relations between an entity pair. We can use this knowledge to prune the training sets of classifier C_i of invalid entity pairs. Relation type signatures overlap but are not identical. Therefore, training set of C_i is different from training set of $C_j, j \neq i$.

2.1 Training

The steps involved in training the aforementioned classifiers are described below.

1. Extract all pairs of candidates (e_a, e_b) that co-occur within a sentence from training documents to form a triple $t = (e_a, e_b, label)$. If e_a and e_b are known to be related by the type r_c, from the relation annotations, we set $label = r_c$. If they are not related, we set $label = NR$. NR is a special label to denote no relation.

2. Add the triple $t = (e_a, e_b, label)$ to the training set of C_i, if (e_a, e_b) satisifies the type signature for relation $r_i, i \in [1, 22]$.

We now have classifier specific training sets, which are sets of triples $t = (e_a, e_b, label)$. To train the classifier, we regard these triples as training examples of class type $label$ and a feature vector constructed for the entity pair (e_a, e_b), as explained in section 2.4.

2.2 Testing

During the test phase, we generate candidate entity pairs from sentences in the test documents. We look up into the relation argument signatures to identify the list of possible relation types for this entity pair. For each such relation type r_i, we test the candidate with the classifier C_i. The entity pair (e_a, e_b) is considered to have the relation type r_i if the predicted label from the classifier C_i is r_i. A consequence of the above approach is that we may predict multiple relation types for a single entity pair in a sentence. This is a limitation of our system, as it is unlikely for a sentence to express multiple relationships between an entity pair.

2.3 Classifier details

The classifiers $C_i, i \in [1, 22]$ are trained as multiclass classifiers. Note that the training set of each classifier C_i may include examples of the form $(e_a, e_b, label), label = r_j$ and $j \neq i$, for the reason that (e_a, e_b) satisfies the type signature for r_i. Therefore, at test time a classifier C_i may classify an entity pair (e_a, e_b) as $r_j, j \neq i$. But we note that r_i is the dominant class for the classifier C_i and other relation types r_j are often under represented during its training. Therefore, we discard predictions r_j from C_i when $j \neq i$. For the entity pair (e_a, e_b) to be included in the final set of predicted relations with the type r_i, we require that the classifier C_i label it as r_i.

We experimented with classifiers from Scikit (Pedregosa et al., 2011). For each relation type, we selected a classifier type between linear kernel SVMs and Multinomial Naive Bayes. This choice was based on performance over development data. We combine the development dataset with training dataset and use it all for training. No parameter tuning was performed.

2.4 Feature Engineering

We developed a set of common lexical, syntactic and dependency parse based features. Relation specific features were also developed. For part of speech tagging and dependency parsing of the text, we used the toolset from Stanford CoreNLP (Manning et al., 2014). These features are described in detail below.

1. Stop word removal: For some relations ("Has Sequence Identical To", "Is Functionally Equivalent To","Regulates Accumulation" and "Regulates Expression") we found that it is beneficial to remove stop words from the sentence.
2. Bag of words: Include all words in the sentence as features, prefixed with "pre","mid" or "post" based on their location with reference to entity mentions in the sentence.
3. Part of Speech (POS): Concatenated sequence of POS tags were extracted separately for words before, after and in the middle of entity mentions in the sentence.
4. Entity features: Entity descriptions and entity types were extracted as features.
5. Dependency path features: We compute the shortest path between the entities in the dependency graph of the sentence and then find the neighboring nodes of the entity mentions along the shortest path. The text (lemma) and POS tags of these neighbors are included as features.
6. Trigger words: For each relation, we designate a few special terms as trigger words and flag their presence as a feature. Trigger words were mainly arrived at by examining the annotation guidelines of the task and a few representative examples.
7. Patterns: A common pattern in text documents is to specify equivalent representations using parenthesis. We find if the two entities are expressed in such a way and include it as a special feature for the relations "Is Functionally Equivalent To" and "Regulates Development Phase".

3 Evaluation

The SeeDev-Binary task objective is to extract all related entity pairs at the document level. The metrics are the standard Precision (P), Recall(R) and F1-score ($\frac{2PR}{P+R}$).

3.1 Dataset

The SeeDev-Binary (Chaix et al., 2016) task provides a corpus of 20 full articles on seed development of *Arabidopsis thaliana*, that have been manually selected by domain experts. This corpus consists of a total of $7,082$ entities and $3,575$ binary relations and is partitioned into training, development and test datasets. Gold standard entity and relation annotations are provided for training and development data and for test data only entity annotations have been released. The given set of 16 entity types are categorized into 7 different entity groups and 22 different relation types are defined. Pre-defined event signatures constrain the types of entity arguments for each relation.

3.2 Results

In the development mode, we used the training dataset for training the relation specific classifiers and predicted the relations over the development dataset. Finally, we trained our classifiers with the full training and development data together. With this system, the predicted relations over the test dataset was submitted to the task. Performance results over the test dataset was made available by the task organizers at the conclusion of the event. These results are detailed in Table 3.2.

4 Discussion

We note that the final relation extraction performance is quite low (36.4%), suggesting that SeeDev-Binary event extraction is a challenging problem. Further, for many event types our system was unable to identify any relation mentions. It is not clear as to why our methods are not effective for these relation types, but it is likely that scarcity of training data is the problem. We observed that our system performed poorly on relation types that have $<$ 100 training samples and has generally succeeded on the rest. It is likely that for these sparsely represented relation types, alternate techniques such as rule based methods might be more successful.

We attempted a few alternate techniques and describe the findings from these approaches below.

4.1 Alternate approaches

1. Two stage approach: We attempted building a first stage general filter that identifies event pairs as "related" or "not related". For this, we grouped all candidate pairs with any of the 22 given relation types into the "positive" class and the rest into the "negative" class in a SVM classifier. In the second stage, we built a multiclass classifier that was to further tune the label of an entity pair from "related" to

Event type	Clasifier used	Metrics on Development data			Metrics on Test data		
		F1	Recall	Prec.	F1	Recall	Prec.
Binds To	SVM	0.269	0.291	0.250	0.262	0.250	0.276
Composes Primary Structure	NB	0.482	0.466	0.500	NA	0	0
Composes Protein Complex	NB	NA	0	0	0.500	0.667	0.400
Exists At Stage	NB	NA	0	0	NA	0	0
Exists In Genotype	SVM	0.248	0.222	0.281	0.354	0.315	0.404
Has Sequence Identical To	SVM	0.336	0.800	0.213	NA	0	0
Interacts With	SVM	0.245	0.218	0.280	0.286	0.241	0.351
Is Functionally Equivalent To	SVM	0.238	0.256	0.222	NA	0	0
Is Involved In Process	SVM	NA	0	0	NA	0	0
Is Localized In	SVM	0.431	0.468	0.400	0.388	0.435	0.351
Is Member Of Family	SVM	0.389	0.545	0.303	0.417	0.523	0.346
Is Protein Domain Of	SVM	0.111	0.068	0.285	0.295	0.419	0.228
Occurs During	NB	NA	0	0	NA	0	0
Occurs In Genotype	SVM	NA	0	0	NA	0	0
Regulates Accumulation	SVM	0.444	0.344	0.625	0.316	0.188	1
Regulates Development Phase	SVM	0.380	0.338	0.434	0.376	0.442	0.327
Regulates Expression	SVM	0.486	0.477	0.495	0.386	0.471	0.327
Regulates Process	NB	0.420	0.513	0.355	0.400	0.394	0.406
Regulates Tissue Development	NB	NA	0	0	NA	0	0
Regulates Molecule Activity	NB	NA	0	0	NA	0	0
Transcribes Or Translates To	NB	0.100	0.076	0.142	NA	0	0
Is Linked To	SVM	NA	0	0	NA	0	0
All Relations	-	0.354	0.360	0.348	0.364	0.386	0.34

Table 1: Results for relation extraction. NB is Multinomial Naive Bayes. Prec is Precision.

one of the 22 relation types. We observed poor performance for the first stage filter and a drop in overall performance.

2. Binary classifiers: We attempted training the classifiers $C_i, i \in [1, 22]$ as binary classifiers, by modifying the triples (e_a, e_b, r_j) to $(e_a, e_b, +)$ if $j == i$ and $(e_a, e_b, -)$ if $j \neq i$. At test time, positive predictions from C_i were inferred as relations r_i. We observed that this approach of combining many subclasses into one negative class reduced precision and hence overall performance.

3. Co-occurrence: A simple approach to relation extraction is to consider all event pairs that occur within a sentence as related. We tried using this cooccurrence strategy for relation types for which SVM or Naive Bayes classifiers did not work effectively. We abandoned this strategy as we observed that the overall F1 score reduced over the development dataset, even as the recall at the relation level improved.

4. Kernel methods: We experimented with the shortest dependency path kernel (Bunescu and Mooney, 2005) and the subset tree kernels (Moschitti, 2006) for classification with SVMs. However their performance was quite low (F1 score < 0.20). It is likely that small training set sizes and multiple entity pairs in most sentences affect the performance of these kernel methods.

5. Dominant class types : In our system we adopted the strategy of only accepting predictions of the dominant class type from each classifier. That is, we filter out predictions of type r_j from classifier C_i when $j \neq i$. This strategy proved very effective when tested over the development dataset. Without this filtering step, we found that our system gets a high recall as expected (0.896) but also too many false positives resulting in low precision (0.027) and F-score (0.053).

True relation type	Predicted relation type														
	NR	BT	CP	EG	HS	IW	IF	IL	IM	IP	RA	RD	RE	RP	TO
NR	NA	21	5	46	43	15	32	33	67	5	6	26	54	157	6
BT	14	7	0	0	0	3	0	0	0	0	0	0	0	0	0
CP	7	0	7	0	1	0	0	0	0	0	0	0	0	0	0
EG	63	0	0	18	0	0	0	0	0	0	0	0	0	0	0
HS	1	0	1	0	16	0	3	0	0	0	0	0	0	0	0
IW	23	0	0	0	0	7	0	0	1	0	0	0	0	0	0
IF	14	0	1	0	14	0	10	0	1	0	0	0	0	0	0
IL	25	0	0	0	0	0	0	22	0	0	0	0	0	0	0
IM	25	0	0	0	1	0	0	0	30	0	0	0	0	0	0
IP	27	0	0	0	0	0	0	0	0	2	0	0	0	0	0
RA	19	0	0	0	0	0	0	0	0	0	10	0	0	0	0
RD	39	0	0	0	0	0	0	0	0	0	0	20	0	0	0
RE	57	0	0	0	0	0	0	0	0	0	0	0	53	0	0
RP	86	0	0	0	0	0	0	0	0	0	0	0	0	92	0
TO	12	0	0	0	0	0	0	0	0	0	0	0	0	0	1

Table 2: Confusion matrix for evaluation over development data using our multiclass classifiers. Rows and columns represent the relations Not Related(NR), Binds To(BT) , Composes Primary Structure(CP) , Exists In Genotype(EG) , Has Sequence Identical To(HS) , Interacts With(IW) , Is Functionally Equivalent To(IF) , Is Localized In(IL) , Is Member Of Family(IM) , Is Protein Domain Of(IP) , Regulates Accumulation(RA) , Regulates Development Phase(RD) , Regulates Expression(RE) , Regulates Process(RP) and Transcribes Or Translates To(TO).

4.2 Error analysis

In Table 4.2 we show the confusion matrix for 16 classifiers of our system, when evaluated over the development dataset. The remaining 6 classifiers were left out as they have 0 predictions and are discussed separately in Section 4.2.1. The entries of the confusion matrix $CM[i, j]$ are the number of test examples whose true type is i and its predicted label is j. From the confusion matrix we see that the primary source of errors is in predicting a relation where there is none or vice versa. Amongst the related entity pairs, the classifier for "Has Sequence Identical To" makes the most errors when the input examples are of type "Is Functionally Equivalent To". Adding more discriminatory features or keywords to discriminate between these two classes is likely to improve performance. Better handling of unrelated entity pairs is likely to be achieved with more syntactic or dependency parse related features, that specifically target the entity mentions in the sentence.

4.2.1 Unsuccessful classifiers

In Table 3.2, the F-score for some of the relation types has been recorded as not available("NA") as our classifiers failed to predict any relations.

Studying the confusion matrix at the classifier level confirms that the classifier did not have enough evidence to detect a relation in many cases. Also, for most of these unsuccessful relation types we observed that the primary class type is underrepresented in their training set. For example, the training sets for the classifier for "Exists At Stage" has $3X$ more examples of type "Regulates Development Phase" than examples of type "Exists At Stage". Better ways of handling class imbalance may improve performance.

5 Conclusion

SeeDev-Binary event extraction was shown to be an important but challenging problem in the BioNLP-Shared Task 2016. This task is also unusual as it calls for the extraction of multiple relation types amongst multiple entity types, often cooccurring in a single sentence. In this paper, we describe our system, which was ranked second with an F1 score of 0.364 in the official results of the task. Our solution was based on a series of supervised classifiers and a rich feature set that contributes to effective relation extraction.

References

Asma Ben Abacha and Pierre Zweigenbaum. 2011. Automatic extraction of semantic relations between medical entities: a rule based approach. *Journal of biomedical semantics*, 2(5):1.

Razvan C. Bunescu and Raymond J. Mooney. 2005. A shortest path dependency kernel for relation extraction. In *Proceedings of the Conference on Human Language Technology and Empirical Methods in Natural Language Processing*, HLT '05, pages 724–731, Stroudsburg, PA, USA. Association for Computational Linguistics.

Estelle Chaix, Bertrand Dubreucq, Abdelhak Fatihi, Dialekti Valsamou, Robert Bossy, Mouhamadou Ba, Louise Delger, Pierre Zweigenbaum, Philippe Bessires, Loc Lepiniec, and Claire Ndellec. 2016. Overview of the regulatory network of plant seed development (seedev) task at the bionlp shared task 2016. In *Proceedings of the 4th BioNLP Shared Task workshop*, Berlin, Germany, August. Association for Computational Linguistics.

Jin-Dong Kim, Tomoko Ohta, Sampo Pyysalo, Yoshinobu Kano, and Jun'ichi Tsujii. 2009. Overview of bionlp'09 shared task on event extraction. In *Proceedings of the Workshop on Current Trends in Biomedical Natural Language Processing: Shared Task*, BioNLP '09, pages 1–9, Stroudsburg, PA, USA. Association for Computational Linguistics.

Jin-Dong Kim, Sampo Pyysalo, Tomoko Ohta, Robert Bossy, Ngan Nguyen, and Jun'ichi Tsujii. 2011. Overview of bionlp shared task 2011. In *Proceedings of the BioNLP Shared Task 2011 Workshop*, BioNLP Shared Task '11, pages 1–6, Stroudsburg, PA, USA. Association for Computational Linguistics.

Haibin Liu, Karin Verspoor, Donald C Comeau, Andrew MacKinlay, and W John Wilbur. 2013. Generalizing an approximate subgraph matching-based system to extract events in molecular biology and cancer genetics. In *Proceedings of the BioNLP Shared Task 2013 Workshop*, pages 76–85.

Christopher D. Manning, Mihai Surdeanu, John Bauer, Jenny Finkel, Steven J. Bethard, and David McClosky. 2014. The Stanford CoreNLP natural language processing toolkit. In *Association for Computational Linguistics (ACL) System Demonstrations*, pages 55–60.

Makoto Miwa, Rune Sætre, Jin-Dong Kim, and Jun'ichi Tsujii. 2010. Event extraction with complex event classification using rich features. *Journal of bioinformatics and computational biology*, 8(01):131–146.

Alessandro Moschitti. 2006. Making tree kernels practical for natural language learning.

Tomoko Ohta, Sampo Pyysalo, Makoto Miwa, and Jun'ichi Tsujii. 2011. Event extraction for dna methylation. *Journal of Biomedical Semantics*, 2(5):1–15.

F. Pedregosa, G. Varoquaux, A. Gramfort, V. Michel, B. Thirion, O. Grisel, M. Blondel, P. Prettenhofer, R. Weiss, V. Dubourg, J. Vanderplas, A. Passos, D. Cournapeau, M. Brucher, M. Perrot, and E. Duchesnay. 2011. Scikit-learn: Machine learning in Python. *Journal of Machine Learning Research*, 12:2825–2830.

Extraction of Regulatory Events using Kernel-based Classifiers and Distant Supervision

Andre Lamurias[1]*, **Miguel J. Rodrigues**[1], **Luka A. Clarke**[2], **Francisco M. Couto**[1],
[1]LaSIGE, Faculdade de Ciências, Universidade de Lisboa, Portugal
[2]BioISI: Biosystems & Integrative Sciences Institute,
Faculdade de Ciências, Universidade de Lisboa, Portugal

Abstract

This paper describes our system to extract binary regulatory relations from text, used to participate in the SeeDev task of BioNLP-ST 2016. Our system was based on machine learning, using support vector machines with a shallow linguistic kernel to identify each type of relation. Additionally, we employed a distant supervised approach to increase the size of the training data. Our submission obtained the third best precision of the SeeDev-binary task. Although the distant supervised approach did not significantly improve the results, we expect that by exploring other techniques to use unlabeled data should lead to better results.

1 Introduction

The SeeDev task of BioNLP-ST 2016 consisted in extracting relations between biomedical named entities on a set of texts about *Arabidopsis thaliana*(Chaix et al., 2016). These texts were manually annotated with entities and relations relevant to seed storage and reserve accumulation. Furthermore, the type of entities that could have a specific role on each type of relation was specified by the organization. There were two subtasks: the first task, binary relation extraction (SeeDev-binary), considered only relations between two arguments; the second, full event extraction, considered relations that could be composed by two to eight arguments. For both tasks, the evaluation criteria used consisted in comparing the type and arguments of each predicted relation to the gold standard. A total of 7 teams participated on this task. The best F-measure achieved was of 0.432,

which is slightly lower than the best scores obtained for the comparable task on the 2013 edition of BioNLP-ST (Cancer Genetics task (Pyysalo et al., 2015): 0.554; Gene Regulation Network task (Bossy et al., 2015): 0.45; GENIA task (Kim et al., 2015): 0.489)

Our team has developed a system for the identification of chemical entities and interactions, based on Conditional Random Fields, kernel methods and domain knowledge. We have also adapted this system to other types of entities such as temporal expressions and clinical events. The SeeDev-binary subtask provided us with an opportunity to test our system on a new domain, which contains more types of entities and relations than the domains we had previously tested on.

We adapted the relation extraction module of our system to the types of relations considered by the SeeDev-binary subtask. For each type of relation, we trained a classifier with the shallow linguistic kernel. We used every sentence containing at least two entities of the types accepted by that relation type. Since there was no ontology readily available for this domain, we were not able to integrate domain knowledge. Alternatively, we experimented a distant supervision approach by using a large number of documents to find sentences containing pairs that were already present on the training corpus. Our system is available at `https://github.com/AndreLamurias/IBEnt`

The following sections describe the main methods used by our system (Section 2), the results obtained with our submission and post-challenge improvements (Section 3), and a discussion about these results (Section 4).

2 Methods

This section describes the methods used by our system. The pre-processing and relation extrac-

---

*Corresponding author: alamurias@lasige.di.fc.ul.pt

Proceedings of the 4th BioNLP Shared Task Workshop, pages 88–92,
Berlin, Germany, August 13, 2016. ©2016 Association for Computational Linguistics

tion steps were already part of our system, implemented for other biomedical domains. For this task, we tested a basic distant supervision approach.

2.1 Pre-processing

The first step of our system consisted in pre-processing the input text using the Genia Sentence Splitter (Sætre et al., 2007) and the Stanford CoreNLP pipeline (Toutanova and Manning, 2000). The latter tokenizes the text into word tokens and extracts the corresponding lemmas and part-of-speech, and named entity tags (proper noun, numerical and temporal entities). We implemented additional tokenization rules to separate words linked by dashes, dots and slashes because biomedical entities may be part of expressions containing these characters.

2.2 Relation extraction

Each of the 22 types of relations has two arguments, and each argument is restricted to a set of entity types specific to each relation type. These restrictions were established by the task organizers. The sentences that satisfied the entity type requirements were considered to train and test a classifier of that relation type. The tokens that comprise the relation arguments were replaced by a generic string in order to reduce the variability of the text. Furthermore, for the types "Has_Sequence_Identical_To" and "Is_Functionally_Equivalent_To", we considered only pairs with the same entity type.

The machine learning algorithm used to train the classifiers was a variation of Support Vector Machines, with the shallow linguistic kernel, as implemented by jSRE (Giuliano et al., 2006). Kernel methods rely on a kernel function which computes the inner product between every instance instead of a specific feature map. This kernel function in particular considers an instance as the sequence of tokens, lemmas, part-of-speech and named entities. The tokens that refer to each argument are identified, while the label of each instance was 0 if the pair was not a relation, or 1 if it was a relation. Each pair of entities that satisfied the argument type restrictions was considered a candidate pair. This kernel has been applied to biomedical text, for the extraction of relations between proteins (Tikk et al., 2010) and chemical compounds (Segura-Bedmar et al., 2011), obtaining positive results. The shallow linguistic kernel is a composite sequence kernel which uses both a local and global context window, which we set at 3 and 4, respectively. These are the only variable parameters of this kernel.

2.3 Distant supervision

The objective of this experiment was to find relations on PubMed abstracts which could increase the size of the training data, and therefore, improve the performance of the system. First, we retrieved the 10,000 most recent abstracts with the MeSH term "arabidopsis" from PubMed. Using the entity annotations from the gold standard, we trained Condition Random Fields (Lafferty et al., 2001) classifiers to recognize each type of entity on the abstracts. We have previously applied this approach to chemical entities, obtaining a F-measure of 0.847 (Lamurias et al., 2015b). We generated lists of the keywords most used in sentences where a relation is described, for each type of relations. To prevent common words from appearing on those lists, we also generated a list of the most used words on the corpus, and removed those words from each list. Our assumption was that if at least two keywords in the list were mentioned in the sentence, then the relation would be true. Since this approach produced mostly negative instances, we excluded some of those to maintain the same positive/negative ratio as the training data. This approach was based on the work of Thomas et al. (2011), where they used various filters to reduce the number of false positives. In this case, we used only instances of the 10 relations types that were least represented in the gold standard. Table 2.3 provides a comparison between the data set obtained with this technique (DS set) and the training set.

3 Results

To classify the test set, we trained with the documents of the gold standard. We present the results of our official submission, as well as the results obtained with the addition of distant supervised sentences (Table 3). More detailed results, as well as the results obtained by the other teams, are available at the task website [1]. After submitting the results, we found that, by mistake, we had trained the classifiers only with the training set. Therefore, we also present the results obtained with the

---

[1] http://2016.bionlp-st.org/tasks/
seedev/seedev-evaluation

		Pairs		Ratio	
Pair type	DS set	Train	DS set	Train
Binds_To	4624	66	0.0449	0.0134
Composes_Primary_Structure	56	32	0.0003	0.0769
Composes_Protein_Complex	16	15	0.0042	0.1172
Exists_At_Stage	400	17	0.0074	0.0499
Is_Involved_In_Process	136	32	0.0127	0.0371
Occurs_In_Genotype	1312	34	0.0194	0.0804
Occurs_During	112	18	0.0032	0.0625
Regulates_Accumulation	5632	65	0.0112	0.0114
Regulates_Molecule_Activity	1664	16	0.0147	0.0015
Regulates_Tissue_Development	704	18	0.0788	0.0060

Table 1: Number of positive pair (Pairs) and positive/negative ratio (Ratio) for each of the relation types considered for the distant supervision approach. DS set refers to the data set generated using distant supervision while Train refers to the training set.

classifiers trained with both training and development sets.

Training	Recall	Precision	F1
Baseline	0.895	0.029	0.056
Train	0.256	0.379	0.306
Train + Dev	0.304	0.341	0.322
Train + Dev + DS	0.366	0.387	0.377

Table 2: SeeDev-binary test set results. Train refers to the training the classifiers with the training set, Dev to the development set and DS to the distant supervision set generated using distant supervision.

Table 3 also contains a baseline that we used during development of the system, to compare the performance of our system to a simple approach. In this case, the simple approach consisted in classifying every pair that satisfied the entity type requirements as a true relation. As expected, this baseline obtained high recall and low precision and F-measure. The reason why the recall is not 1 is because we only considered pairs of entities from the same sentence. This way, the recall of the baseline (0.895) is the maximum recall we could have obtained with our approach. We observed that with our system, the results obtained were better both in terms of precision and F-measure.

The main difference between training with just the training set and using both training and development was in the recall obtained. By increasing the number of training instances, the classifier was able to correctly identify more relations. Although it also decreased the precision, the difference in terms of F-measure was positive.

Using the distant supervision approach, we were able to use 6947 sentences as an additional data set (DS set). This approach improved the F-measure by 0.055, due to an increase in recall and precision.

4 Discussion

This task was a challenge for our system since it required the identification of 22 types of relations, while previously the system was tested only on one specific type of relation While we could optimize the system for one type of relation with domain knowledge, in this case we had to use a generic approach to various types.

Comparing with the other participants, our F-measure was the 5th best of the 7 participating teams, 0.126 points below the best. In terms of precision, our team was the 3rd best, 0.154 below the best. Our submitted results had higher precision because we used only the gold standard annotations to train the classifiers. This way, the output of the classifiers tended to be closer to the training corpora.

4.1 Error Analysis

In order to fairly compare our results with the other teams, we discuss only the errors of our official submission. There was a wide range of F-measure values within the different types of relations. The types "Has_Sequence_Identical_To" and "Is_Functionally_Equivalent_To" had a F-measure of 0.708 and 0.646, respectively. These types obtain much higher scores possibly because the entity types of the two arguments had to be the same, reducing the number of candidate pairs. The most difficult relations were the ones less represented in the training data, such as "Is_Involved_In_Process" and "Is_Linked_To". In the case of the first type, no team was able to identify one of the 12 relation instances present in the test corpus, while with the second type, only one team was able to identify some relations. These results show that the performance of the techniques used for this task are dependent on the annotations of the training data.

Regarding the contribution of the distant supervision approach, we observed that the system predicted fewer relations of the less frequent relation types. Since we labeled each pair of entities automatically, it is possible that some relations were mislabeled. However, since we maintained the same positive/negative ratio as the training set (Table 2.3), this approach provided mostly negative instances.

4.2 Future Work

We intend to explore other techniques to use unlabeled data for distant supervision. A technique that has improved results on other domains consists of using a knowledge base to restrict which entities could constitute a relation (Bunescu and Mooney, 2007). By combining the knowledge base with the keyword based filter, we should obtain a set of instances with a high probability of being correctly labeled. These instances should then improve the quality of the classifiers by providing other ways to express a relation, and reduce the number of incorrect annotations.

Another technique to explore consists in applying semantic similarity measures (Couto and Pinto, 2013) to check if two entities are semantically related and therefore could constitute a relation (Lamurias et al., 2015a). Additionally, we intend to apply our distant supervision approach to improve the results of our biomedical question&answering system (WS4A) that participated in the BioASQ 2016 challenge(Rodrigues et al., 2016).

Acknowledgments

This work was supported by the Fundação para a Ciência e a Tecnologia (https://www.fct.mctes.pt/) through the PhD grant PD/BD/106083/2015 and LaSIGE Unit Strategic Project, ref. UID/CEC/00408/2013 (LaSIGE).

References

[Bossy et al.2015] Robert Bossy, Wiktoria Golik, Zorana Ratkovic, Dialekti Valsamou, Philippe Bessières, and Claire Nédellec. 2015. Overview of the gene regulation network and the bacteria biotope tasks in bionlp'13 shared task. *BMC bioinformatics*, 16(Suppl 10):S1.

[Bunescu and Mooney2007] Razvan Bunescu and Raymond Mooney. 2007. Learning to extract relations from the web using minimal supervision. In *Annual meeting-association for Computational Linguistics*, volume 45, page 576.

[Chaix et al.2016] Estelle Chaix, Bertrand Dubreucq, Abdelhak Fatihi, Dialekti Valsamou, Robert Bossy, Mouhamadou Ba, Louise Delger, Pierre Zweigenbaum, Philippe Bessires, Loc Lepiniec, and Claire Ndellec. 2016. Overview of the regulatory network of plant seed development (seedev) task at the bionlp shared task 2016. In *Proceedings of the 4th BioNLP Shared Task workshop*, Berlin, Germany, August. Association for Computational Linguistics.

[Couto and Pinto2013] Francisco M Couto and H Sofia Pinto. 2013. The next generation of similarity measures that fully explore the semantics in biomedical ontologies. *Journal of bioinformatics and computational biology*, 11(05):1371001.

[Giuliano et al.2006] Claudio Giuliano, Alberto Lavelli, and Lorenza Romano. 2006. Exploiting shallow linguistic information for relation extraction from biomedical literature. In *EACL*, volume 18, pages 401–408. Citeseer.

[Kim et al.2015] Jin-Dong Kim, Jung-jae Kim, Xu Han, and Dietrich Rebholz-Schuhmann. 2015. Extending the evaluation of genia event task toward knowledge base construction and comparison to gene regulation ontology task. *BMC bioinformatics*, 16(Suppl 10):S3.

[Lafferty et al.2001] John Lafferty, Andrew McCallum, and Fernando CN Pereira. 2001. Conditional random fields: Probabilistic models for segmenting and labeling sequence data.

[Lamurias et al.2015a] Andre Lamurias, João D Ferreira, and Francisco M Couto. 2015a. Improving chemical entity recognition through h-index based semantic similarity. *Journal of cheminformatics*, 7(1):1.

[Lamurias et al.2015b] Andre Lamurias, Manuel Lobo, Marta Antunes, Luka A Clarke, and Francisco M Couto. 2015b. Identifying chemical entities in patents using brown clustering and semantic similarity. In *5th BioCreative Challenge Evaluation*.

[Pyysalo et al.2015] Sampo Pyysalo, Tomoko Ohta, Rafal Rak, Andrew Rowley, Hong-Woo Chun, Sung-Jae Jung, Sung-Pil Choi, Jun'ichi Tsujii, and Sophia Ananiadou. 2015. Overview of the cancer genetics and pathway curation tasks of bionlp shared task 2013. *BMC bioinformatics*, 16(Suppl 10):S2.

[Rodrigues et al.2016] Miguel J. Rodrigues, Miguel Falé, Andre Lamurias, and Francisco M. Couto. 2016. WS4A: a biomedical question and answering system based on public web services and ontologies. Poster session at the 4th BioASQ Workshop.

[Sætre et al.2007] Rune Sætre, Kazuhiro Yoshida, Akane Yakushiji, Yusuke Miyao, Yuichiro Matsubayashi, and Tomoko Ohta. 2007. AKANE system: protein-protein interaction pairs in BioCreAtIvE2 challenge, PPI-IPS subtask. In *Proceedings of the Second BioCreative Challenge Workshop*, pages 209–212.

[Segura-Bedmar et al.2011] Isabel Segura-Bedmar, Paloma Martinez, and Cesar de Pablo-Sánchez. 2011. Using a shallow linguistic kernel for drug–drug interaction extraction. *Journal of biomedical informatics*, 44(5):789–804.

[Thomas et al.2011] Philippe Thomas, Illés Solt, Roman Klinger, and Ulf Leser. 2011. Learning to extract protein–protein interactions using distant supervision. *ROBUS 2011*, page 25.

[Tikk et al.2010] Domonkos Tikk, Philippe Thomas, Peter Palaga, Jörg Hakenberg, and Ulf Leser. 2010. A comprehensive benchmark of kernel methods to extract protein–protein interactions from literature. *PLoS Comput Biol*, 6(7):e1000837.

[Toutanova and Manning2000] Kristina Toutanova and Christopher D Manning. 2000. Enriching the knowledge sources used in a maximum entropy part-of-speech tagger. In *Proceedings of the 2000 Joint SIGDAT conference on Empirical methods in natural language processing and very large corpora: held in conjunction with the 38th Annual Meeting of the Association for Computational Linguistics-Volume 13*, pages 63–70. Association for Computational Linguistics.

DUTIR in BioNLP-ST 2016: Utilizing Convolutional Network and Distributed Representation to Extract Complicate Relations

Honglei Li, Jianhai Zhang, Jian Wang, Hongfei Lin, Zhihao Yang
School of Computer Science and Technology
Dalian University of Technology
116024 Dalian, China
201081023@mail.dlut.edu.cn
jianhai0527@mail.dlut.edu.cn
wangjian@dlut.edu.cn
hflin@dlut.edu.cn
yangzh@dlut.edu.cn

Abstract

We participate in the two event extraction tasks of BioNLP 2016 Shared Task: binary relation extraction of SeeDev task and localization relations extraction of Bacteria Biotope task. Convolutional neural network (CNN) is employed to model the sentences by convolution and max-pooling operation from raw input with word embedding. Then, full connected neural network is used to learn senior and significant features automatically. The proposed model mainly contains two modules: distributive semantic representation building, such as word embedding, POS embedding, distance embedding and entity type embedding, and CNN model training. The results with F-score of 0.370 and 0.478 in our participant tasks, which were evaluated on the test data set, show that our proposed method contributes to binary relation extraction effectively and can reduce the impact of artificial feature engineering through automatically feature learning.

1 Introduction

Information extraction devotes to finding useful data and hidden knowledge for researchers from amounts of texts. With the demands of rapidly and accurately locating key issues about life and biology increasing, bio-IE appears timely and has attracted more and more researchers to address this question (Krallinger et al., 2005; Zweigenbaum et al., 2007). Much progress has been made in named entity identification, protein-protein relations classification (Blaschke et al., 1999) and drug-drug interaction extraction (Rodrigues et al., 2008). Furthermore, fine-grained information extraction in biology, in particular event extraction has entered the spotlight of people and, appeared many meaningful and challenge tasks for event extraction, which can gather the community-wide efforts and contribute to the development of biology information extraction (Kim et al., 2009; Kim et al., 2011; Nédellec et al., 2013).

The BioNLP Shared Task series (Kim et al., 2009; Kim et al., 2011; Nédellec et al., 2013) is a representative for biomolecular event extraction, which has been held four times including this year. The topics of the series range from fine-grained extraction, generalization to knowledge base construction. In addition, the scope that this task involved has become much broader at each edition. For example, BioNLP-ST 2013 (Nédellec et al., 2013) covers many new hot topics compared to the previous editions, such as Cancer Genetics, Pathway Curation and Gene Regulation Network in Bacteria.

BioNLP-ST 2016 further broadens the scope of the text-mining application domains in biology by introducing a new issue on seed development, named the issue as the SeeDev task. The development of the seed is a critical issue in agriculture and presents an opportunity for the community to contribute the common efforts in bio-IE. The other task, Bacteria Biotope of the BioNLP-St'13 expands on the previous editions by replacing the Web Pages with scientist papers abstracts to organize the corpus, which is much closer to the actual needs of detailed and scientific information for biologists. The third task focuses on the Genia corpus as previous edition, but gives more emphasis in the contribution from any aspect of knowledge extraction, which is an open question to participants.

We focus on the two events extraction subtasks of BioNLP 2016 Shared Task: binary rela-

93

Proceedings of the 4th BioNLP Shared Task Workshop, pages 93–100,
Berlin, Germany, August 13, 2016. ©2016 Association for Computational Linguistics

tion extraction of SeeDev task and localization relations extraction of Bacteria Biotope task. Both tasks broaden the scope of fine-grained information extraction in biology, and contribute to the development of the actual application in text mining.

The SeeDev task has not been introduced in the previous BioNLP-ST and aims at exploring the knowledge of the molecular network underlying the regulation of seed development. The SeeDev task is similar to the GRN (Gene Regulation Network in Bacteria) task in BioNLP'13, aiming at extracting a regulation network that links and integrates a variety of molecular (Bossy et al., 2013) or processes interactions between entities. Therefore, the superior systems from the GRN can give us some useful heuristics. Five systems participated in GRN and all systems applied machine learning algorithms with many different resources of information and preprocessing in BioNLP'13. Lots of features, such as linguistic features, semantic and syntactic information between two entities, were added into these systems. However, they implemented different ML algorithms, including SVM, CRF and KNN (Bossy et al., 2013). For example, Provoost (2013) employed a basic Support Vector Machine framework and focused more on the domain of feature definition and exploration. They achieved an F-score of 0.313, standing on second place in GRN task of BioNLP'13. IRISA system (Claveau, 2013) emphasized the similarity between the known instances and the closest known examples based on K-Nearest Neighbor algorithm.

Bacteria Biotope task in the BioNLP-ST 2016, our second participation, was the third edition that focuses on extracting localization relations between bacteria and their habitats from scientific papers abstracts. Many systems had contributed their efforts to the task in the precious editions. Boun system (Karadeniz et al., 2013) used the shallow linguistic knowledge of the corpus to implement the prediction based on previously defined syntactic rules and discourse-based rules, coming the F-score of 0.27. The Alvis system (Ratkovic et al., 2011) also employed hand-designed patterns to detect the relations between bacteria and habitat with the linguistic and lexical knowledge. UTurku and JAIST (Karadeniz et al., 2013) systems in BioNLP'11 explored different approaches from the above mentioned and regarded the binary event extraction as a classification problem, thus applying machine learning methods. In BioNLP'13, TEES-

2.1 and IRISA (Bossy et al., 2013) also employed the same idea to this question, and achieved the state-of-the-art results with F-score of 0.42 and 0.40, respectively, which were much higher than the two hand-designed rules methods: LIMSI and Boun.

Most of systems delivered their good ideas and achieved the better results for these tasks in BioNLP-ST, which have positively promoted the development of biology information extraction. So, it is an opportunity for researchers to apply various approaches and new ideas to these tasks. Over recent years, the landscape of Convolutional Neural Network (CNN) has been obviously prosperous and pushed forward through the expansion of actual application of various fields. The introduction of convolution layers and pooling layers in CNN has helped to improve the performance of features automatically learnt in networks. Therefore, in our work, we explore the CNN to learn features automatically for the two binary relations extraction tasks, significantly differenced from previous systems in BioNLP-ST.

2 Method

The tasks of SeeDev-binary and BB-event both can be treated as binary relation extraction which specifics whether there is interaction between two entities. In relation extraction, the semantic and syntactic information for sentence act as a significant role. Traditional method usually need to design and extract complex features from sentence based on domain-specific knowledge, such as tree kernel and graph kernel, to model the sentences. As a result, this will lead to much lower ability of generation for corpus dependent. Consequently, instead of complicate hand-designed feature engineering, we employ convolutional neural network, also called CNN, to model the sentences by convolution and max-pooling operation from raw input with word embedding and full connected neural network to learn senior features automatically. Furthermore, we employ POS embedding to enrich the semantic information of words, distance embedding to capture the information of relative distance between the entities and entity type embedding as the supplement features of the sentence. All the feature embedding is combined to build final distributive semantic representation which is fed to convolutional neural network.

As described in Fig.1, the proposed model mainly contains two modules: distributive semantic representation building, such as word embedding, POS embedding, distance embedding and entity type embedding, and CNN model training. In the next parts, we will introduce more details.

2.1 Build Distributive Semantic Representation

Traditional one-hot representation, which is employed by mostly machine learning methods, can vectorize the text and plays an important role. However, it can result in the problems of semantic gap and dimension disaster which restrict its application. Consequently, in our proposed method, we employ distributive semantic representation, proposed by Hinton (1986) at first, as the feature representation of the model. And then, we exploit the advantage of convolutional neural network at modeling the sentences to learn sentence-level representation from raw input. The distributive semantic representation is built as follows. For simply definition, we assume $S = E_1 W_1 W_2 W_3 \dots W_n E_2$ as the word sequence between two entities in one sentence, where E_1, E_2 stand for the entities and $W_1 \dots W_n$ stand for the words between two entities.

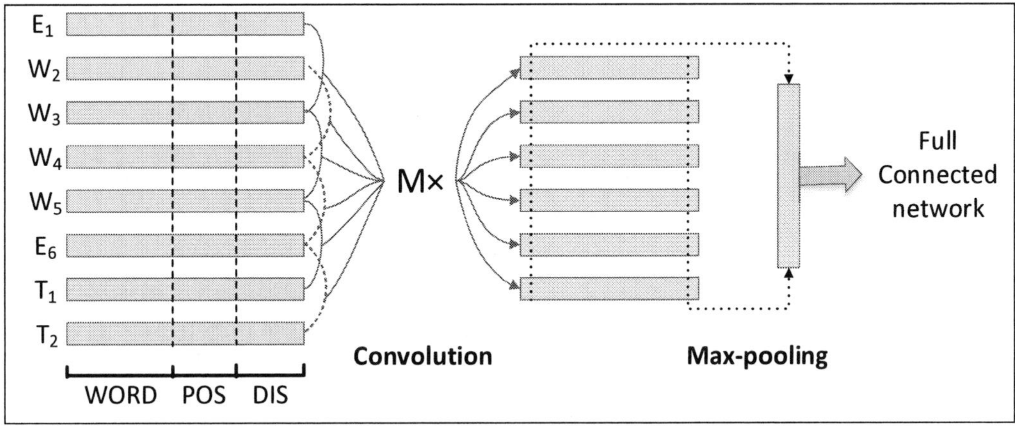

Figure 1: The model of convolutional neural network with distributive vector

2.1.1 Word Embedding

Instead of traditional one-hot representation, we utilize the distributive semantic representation of words for solving the problem of dimension disaster and semantic gap. Firstly, we employ word2vec tool, which can effectively learn distributive representation of words from massive and unlabeled data, to train word embedding from massive available Pubmed abstracts. The embedding with low dimension and realistic value contains rich semantic information and can be treated as feature representation of words instead of one-hot.

Inspired by language model, we employ the contexts of two entities to predict the relation type. In our experiments, the contexts are expressed by the words between two entities in one sentence. Then, the word sequence is transformed into word embedding matrix by looking up the word embedding table. The word embedding matrix can be treated as local feature of the sentence and fed to CNN model to learn global feature which can contribute to the relation identification. The word embedding matrix is represented as follows:

$$LT_W(S) = \left[\langle W \rangle_{E_1}, \langle W \rangle_{W_1}, \langle W \rangle_{W_2}, \dots, \langle W \rangle_{W_n}, \langle W \rangle_{E_2} \right]$$

Where $W \in \mathbb{R}^{|\mathcal{D}| \times dim}$ ($|\mathcal{D}|$ is the size of dictionary and dim is the dimension of word embedding) is the word embedding table trained by word2vec with Pubmed abstracts and fine-tuned while training.

2.1.2 Entity Type Embedding

Through analyzing the dataset, we observe that different entities with different types have different probability to interact with each other if the entity type satisfies the relation constraints. Consequently, entity type of two entities is an import factor for predicting the relation type. In our model, entity types are treated as the extra features of the relation and the supplement of word sequence. $\langle W^T\rangle_{type(E_1)}$, $\langle W^T\rangle_{type(E_2)}$ are added as the extra features of the relation:

$$LT_{W,W^T}(S) =$$
$$[\langle W\rangle_{E_1}, \langle W\rangle_{W_1}, \dots, \langle W\rangle_{E_2}, \langle W^T\rangle_{type(E_1)}, \langle W^T\rangle_{type(E_2)}]$$

Where $W^T \in \mathbb{R}^{|\mathcal{D}_T|\times dim}$ is type embedding which is randomly initialized by random sampling from the uniform distribution ([-0.25, 0.25]). $type(\cdot)$ stands for the entity type. $\mathcal{D}_T$ is the dictionary of entity types.

2.1.3　POS Embedding

Word semantics usually have several aspects containing similarity, POS (part-of-speech) and so on. For enriching the semantic representation of each word, POS embedding is introduced as the supplement of word embedding:

$$LT_{W^p}(S) =$$
$$[\langle W^p\rangle_{p(E_1)}, \langle W^p\rangle_{p(W_1)}, \dots, \langle W^p\rangle_{p(W_n)}, \langle W^p\rangle_{p(E_2)}, \mathbb{0}, \mathbb{0}]$$

We denote $W^p \in \mathbb{R}^{|\mathcal{D}_p|\times dim_p}$ as the POS embedding which is randomly initialized as well as type embedding, where $\mathcal{D}_p$ is the size of POS dictionary and, dim_p, a hyper-parameter, is the dimension of POS embedding. We set $dim_p = 5$ through trying different configuration. Zero vector ($\mathbb{0}$) is used to pad the sentence.

2.1.4　Distance Embedding

In relation classification tasks, distance information usually plays an important role. Distance can capture the relative position between two entities. As shown in followed formulas, $LT_{W^d}(S)_1$ stands for the relative distance between words and the first entity, and $LT_{W^d}(S)_2$ stands for the relative distance between words and the second entity.

$$LT_{W^d}(S)_1 =$$
$$[\langle W^d\rangle_{d(E_1,E_1)}, \dots, \langle W^d\rangle_{d(W_n,E_1)}, \langle W^d\rangle_{d(E_2,E_1)}, \mathbb{0}, \mathbb{0}]$$

$$LT_{W^d}(S)_2 =$$
$$[\langle W^d\rangle_{d(E_1,E_2)}, \dots, \langle W^d\rangle_{d(W_n,E_2)}, \langle W^d\rangle_{d(E_2,E_2)}, \mathbb{0}, \mathbb{0}]$$

Where $W^d \in \mathbb{R}^{|\mathcal{D}_d|\times dim_d}$ stands for the distance embedding and $|\mathcal{D}_d|$ is the number of different distances. The embedding is randomly initialized and fine-tuned while training. We set $dim_d = 5$ through trying different confiuration. Zero vector ($\mathbb{0}$) is also used to pad the sentence.

As shown in followed formula, the final distributive semantic representation is acquired by joining the word embedding, type embedding, POS embedding and distance embedding.

$$\varphi(S) = \begin{bmatrix} LT_{W,W^T}(S) \\ LT_{W^p}(S) \\ LT_{W^d}(S)_1 \\ LT_{W^d}(S)_2 \end{bmatrix}$$

2.2　Model Training and Parameters Tuning

After building the distributive semantic representation of relation, we employ convolution and max-pooling to learn the global feature representation from raw input. The detailed computation procedure is described as follows.

$$\langle f\rangle_t = f(W \cdot \varphi(S) + b)$$

$$\langle h\rangle = \max_t \langle f\rangle_t$$

Where W is the convolution filter, it extracts local features from given window of word sequence. $\langle h\rangle$ can be treated as the global feature representation learned from raw distributive representation $\varphi(S)$ and be fed to the full connection layer to learn hidden and senior features.

As we all know, convolutional neural network is a model with vast computation cost. Consequently, we implement the CNN model with theano (Bergstra et al., 2010; Bastien et al., 2012) and run in GPU kernels for accelerating the training procedure. As a result, it takes about half hour to train a CNN model. Meanwhile, we make some modifications in our model for achieving more significant experiment results. In the convolutional layer, we make use of multiple convolution kernels with different window size for capturing sentence features from different views. In the full connection layer, we modified the network with dropout (Srivastava et al., 2014) which is a much simple and efficient method to prevent the problem of overfitting. The dropout network can prevent the co-adaption between the nodes through randomly dropping some nodes or make them not work. Learning rate is the most important hyper-parameter in deep learning. Consequently, we employ Adadelta (Zeiler, 2012) an adaptive learning rate method, to automatically adapt the learning rate instead of configuring it manually. Finally, we empirically search for the reasonable combination of all the hyper-parameters and tune in development dataset. The optimal parameters of CNN model are described in Table 1.

hyper-parameter	value
Word embedding	50
filter	1800
window	[3,5,7]
layer	3
dropout	0.3
batch	128

Table 1: The parameters of CNN model

3 Results and Discussions

This section presents our results on the SeeDev-binary and BB-event tasks respectively.

3.1 The results of SeeDev-binary task

The SeeDev-binary task datasets contains three parts, namely the training set, development set and test set respectively, which are totally 87 segments from 20 full articles on seed development of Arabidopsis thaliana. The task defines 17 different types of entities and 22 different types of binary relations. Table 2 shows the detailed distribution of data.

	#	Train	Dev	Test
Segments	87	39	19	29
Entities	7082	3259	1607	2216
Binary relations	3575	1628	819	1128

Table 2: Detailed statistics of SeeDev-binary task corpus

We aim at extracting the relation between the two target entities and reducing the participa-

tion of hand-designed feature engineering by using our proposed model. Table 3 lists the results of our method on the development and test datasets for SeeDev-binary task. The first two lines are the systems with the two best F-score in official results in BioNLP-ST 2016.

Our method achieved the F-score of 0.368 and 0.370 on the development set and test set, respectively. Compared to the official results from different systems, we stood the similar place with the second best system UniMelb which achieved the F-score of 0.364. It demonstrates that our proposed method has a good performance on binary relations extraction.

In previous methods to binary relations classification, more systems prefer to rules-based or feature engineering methods. However, we employ a different idea, which utilizes the advantages of distributive semantic representation and the CNN model. From the detailed results in Table 4, we can find that the proposed model is of benefit to SeeDev binary task. Moreover, the better recall than precision is achieved on the test datasets. In Table 4, four relations, such as "Occurs_In_Genotype", and "Regulates_Molecule_Activity", are not identified by the system, which may be a reason that the size of these relations in corpus is very small.

Methods	Recall	Precision	F-score
LitWay	0.448	0.417	0.432
UniMelb	0.386	0.345	0.364
Our method (on dev set)	0.396	0.344	0.368
Our method (on test set)	0.417	0.333	0.370

Table 3: Results of our method on the development and test data sets for SeeDev-binary task

	Binary relation type	Dev data set	Test data set
		R/P/F-score	R/P/F-score
When and Where	Exists_In_Genotype	0.506/0.273/0.355	0.520/0.361/0.426
	Occurs_In_Genotype	0.000/0.000/0.000	0.000/0.000/0.000
	Exists_At_Stage	0.125/0.100/0.111	0.100/0.045/0.063
	Occurs_During	0.200/0.333/0.250	0.083/0.143/0.105
	Is_Localized_In	0.426/0.253/0.318	0.290/0.231/0.257
Function	Is_Involved_In_Process	0.000/0.000/0.000	0.000/0.000/0.000
	Transcribes_Or_Translates_To	0.154/0.286/0.200	0.313/0.208/0.250
	Is_Functionally_Equivalent_To	0.575/0.821/0.677	0.636/0.745/0.686
Regulation	Regulates_Accumulation	0.103/0.231/0.143	0.125/0.100/0.111
	Regulates_Development_Phase	0.119/0.206/0.151	0.221/0.218/0.219
	Regulates_Expression	0.451/0.485/0.467	0.370/0.307/0.336
	Regulates_Molecule_Activity	0.000/0.000/0.000	0.000/0.000/0.000
	Regulates_Process	0.693/0.363/0.476	0.613/0.357/0.451

	Regulates_Tissue_Development	0.000/0.000/0.000	0.000/0.000/0.000
Composition and Membership	Composes_Primary_Structure	0.200/0.500/0.286	0.563/0.750/0.643
	Composes_Protein_Complex	0.000/0.000/0.000	0.667/0.067/0.121
	Is_Protein_Domain_Of	0.172/0.278/0.213	0.129/0.400/0.195
	Is_Member_Of_Family	0.364/0.308/0.333	0.547/0.338/0.418
	Has_Sequence_Identical_To	0.613/0.905/0.731	0.730/0.852/0.786
Interaction	Interacts_With	0.281/0.500/0.360	0.019/0.500/0.036
	Binds_To	0.208/0.227/0.217	0.188/0.240/0.211
Other	Is_Linked_To	0.087/0.133/0.105	0.350/0.350/0.350
	=[ALL RELATIONS]=	0.396/0.344/0.368	0.417/0.333/0.370

Table 4: Detailed results of our method on the development and test data sets for SeeDev-binary task

3.2 The results of BB-event task

For localization relations extraction of Bacteria Biotope task, we also use our proposed system to evaluate the performance. Table 5 shows the results on the development and test datasets. The F-score of 0.478 on test dataset suggest that the proposed method has positive effects on identifying the binary relation. However, the recall on the test dataset is lower than the precision, which may be overfitting on training data. The F-score of 0.499 on the development data set achieve better performance than that on test data set.

The prediction of location relations remains many challenges. First, high diversity of bacteria and locations increases the difficult of the correct pairing. Second, cross-sentences relations caused by coreferences usually are ignored by most system due to complexity and difficulties. In our system, we only considered the relations in one sentence, which many relations in cross sentences were ignored and might cause some reduce on the performance.

Data set	Recall	Precision	F-score
Dev	0.561	0.449	0.499
Test	0.397	0.600	0.478

Table 5: Results of our method on the development and test data sets for BB-event task

From above analysis, the cross-sentences relations extraction is a big challenge, due to much coreferences relations and increasing negative examples. We conduct another experiment to extract relations at the documental level, but not considering the coreferences resolution. Table 6 shows the evaluated results of our method on the development set and test sets at the documental level and sentence level.

At the documental level, the F-score has an about 2% increase on development dataset, while the F-score increases by 6% on test dataset. It

may be because the distribution of relations on two datasets has large different, which there are more cross-sentence relations on test dataset than development dataset. Furthermore, Table 7 shows the statistics of positive and negative examples on training data and development data at the two levels. (It is not nearly possible to have relations between two candidate entities if their distance is too large. Therefore, we remove the candidate examples if the distance is larger than 60.) We can find that, the ratio between positive and negative examples at the documental level is significantly higher than that at the sentence level. The imbalance between positive and negative examples can significantly influence the performance of models. Therefore, we should devote more techniques and good designs to cross-sentences relation extraction.

Models	Recall	Precision	F-score
CNN-Doc (on dev set)	0.552	0.496	0.523
CNN-Sen (on dev set)	0.561	0.449	0.499
CNN-Doc (on test set)	0.563	0.515	0.538
CNN-Sen (on test set)	0.397	0.600	0.478

Table 6: Results of our method on the development and test data sets for BB-event task

Models	#Positive examples	#Negative examples	Ratio
Doc-level (on train set)	16%(298)	84%(1525)	5.1
Sen-level (on train set)	45%(227)	55%(275)	1.2
Doc-level (on dev set)	13%(210)	87%(1462)	6.9
Sen-level (on dev set)	32%(165)	68%(348)	2.1

Table 7: Statistics of positive and negative examples on training data and development data at the documental and sentence levels for BB-event task (ratio = #negative examples / #positive examples).

We conduct another experiment on SVM[1] to analysis the superiority of CNN model compared with SVM model. Each raw input into the SVM and CNN models is same, which contains words between two candidate entities, distance between two candidate entities, and the types of two candidate entities. Then, the raw input for SVM is represented traditional one-hot features, and the raw input for CNN is represented by distributed representation. In Table 8, we compared the two models. F-score of using CNN model is higher than that using SVM model on two data sets, which shows that the effectiveness of using CNN model and distributed representation.

Models	Recall	Precision	F-score
SVM (on dev set)	0.459	0.490	0.474
CNN (on dev set)	0.561	0.449	0.499
SVM (on test set)	0.336	0.594	0.429
CNN (on test set)	0.397	0.600	0.478

Table 8: Results of using SVM and CNN models on the development and test data sets for BB-event task

4 Conclusions

Instead of complicate hand-designed feature engineering, we employed the distributed semantic representation and CNN model to extract binary relations between entities. SeeDev-binary task and BB-event task are regarded as classification problems. And then, Word embedding, POS embedding, distance embedding and entity type embedding, which contain rich semantic knowledge, are built to be fed into Convolutional neural network and to learn the inner relationship between candidate entities. The results with F-score of 0.370 and 0.478 in our participant tasks, which were evaluated on the test data set with online evaluation[2] show that our proposed method has been contributed to binary relation extraction.

Only using embedding of original words fed into CNN, may be not sufficient for understanding the hidden information among words. Therefore, in our future work, we will still concentrate more on the building of rich distributed semantic embedding and construct a better representation with human knowledge for CNN model. Furthermore, we will also explore various neural networks with multi-layer architectures, such as RNN, to address binary relation or event extraction.

5 Acknowledge

This work is supported by the grants from the National Natural Science Foundation of China (No. 61572098, 61572102, 61272373 and 61300088), Trans-Century Training Program Foundation for the Talents by the Ministry of Education of China (NCET-13-0084) and the Fundamental Research Funds for the Central Universities (No. DUT13JB09).

References

Blaschke C, Andrade M A, Ouzounis C A, et al. Automatic extraction of biological information from scientific text: protein-protein interactions[C]//Ismb. 1999, 7: 60-67.

Bergstra J, Breuleux O, Bastien F, et al. Theano: a CPU and GPU math expression compiler[C]//Proceedings of the Python for scientific computing conference (SciPy). 2010, 4: 3.

Bastien F, Lamblin P, Pascanu R, et al. Theano: new features and speed improvements[J]. arXiv preprint arXiv:1211.5590, 2012.

Bossy R, Bessières P, Nédellec C. Bionlp shared task 2013–an overview of the genic regulation network task[J]. ACL 2013, 2013: 153.

Bossy R, Golik W, Ratkovic Z, et al. BioNLP Shared Task 2013–an overview of the bacteria biotope task[C]//Proceedings of the BioNLP Shared Task 2013 Workshop. 2013: 161-169.

Claveau V. IRISA participation to BioNLP-ST 2013: lazy-learning and information retrieval for information extraction tasks[C]//BioNLP Workshop, colocated with ACL 2013. 2013: 188-196.

Hinton G E. Learning distributed representations of concepts[C]//Proceedings of the eighth annual conference of the cognitive science society. 1986, 1: 12.

Krallinger M, Erhardt R A A, Valencia A. Text-mining approaches in molecular biology and bi-

---

[1] http://www.cs.cornell.edu/People/tj/svm_light/
[2] http://2016.bionlp-st.org/tasks/seedev/seedev-evaluation

omedicine[J]. Drug discovery today, 2005, 10(6): 439-445.

Kim J D, Ohta T, Pyysalo S, et al. Overview of BioNLP'09 shared task on event extraction[C]//Proceedings of the Workshop on Current Trends in Biomedical Natural Language Processing: Shared Task. Association for Computational Linguistics, 2009: 1-9.

Kim J D, Pyysalo S, Ohta T, et al. Overview of BioNLP shared task 2011[C]//Proceedings of the BioNLP Shared Task 2011 Workshop. Association for Computational Linguistics, 2011: 1-6.

Karadeniz I, Ozgür A. Bacteria biotope detection, ontology-based normalization, and relation extraction using syntactic rules[C]//Proceedings of the BioNLP Shared Task 2013 Workshop. 2013: 170-177.

Nédellec C, Bossy R, Kim J D, et al. Overview of BioNLP shared task 2013[C]//Proceedings of the BioNLP Shared Task 2013 Workshop. 2013: 1-7.

Provoost T, Moens M F. Detecting relations in the gene regulation network[C]//Proceedings of BioNLP shared task 2013 workshop: the Genia event extraction shared task. 2013: 135-138.

Rodrigues A D (Ed). Drug-drug interactions. CRC Press. 2008.

Ratkovic Z, Golik W, Warnier P, et al. BioNLP 2011 task bacteria biotope: the Alvis system[C]//Proceedings of the BioNLP Shared Task 2011 Workshop. Association for Computational Linguistics, 2011: 102-111.

Srivastava N, Hinton G, Krizhevsky A, et al. Dropout: A simple way to prevent neural networks from overfitting[J]. The Journal of Machine Learning Research, 2014, 15(1): 1929-1958.

Zweigenbaum P, Demner-Fushman D, Yu H, Cohen KB. Frontiers of biomedical text mining: current progress. Brief Bioinform. 2007, 8:358-375.

Zeiler M D. ADADELTA: an adaptive learning rate method[J]. arXiv preprint arXiv:1212.5701, 2012.

Extracting Biomedical Event Using Feature Selection and Word Representation

Xinyu He, Lishuang Li*, Jieqiong Zheng, Meiyue Qin
School of Computer Science and Technology
Dalian University of Technology
116023 Dalian, China
`lilishuang314@163.com hexinyu123@163.com`

Abstract

We participate in the BB3 and GE4 tasks of BioNLP-ST 2016. In the BB3 task, we adopt word representation methods to improve the feature-based Biomedical Event Extraction System, and take the 4th place. In the GE4 task, based on the Uturku system, a two-stage method is proposed for trigger detection, which divides trigger detection into recognition stage and classification stage, using different features in each stage. In the edge detection, we adopt Passive-aggressive (PA) online algorithm, then we constitute events by post-processing of TEES.

1 Method

In the BB3 task, we improve the performance of the biomedical event extraction by word representation methods, which include distributed word representation, and Brown clusters representation. The framework of the proposed system includes input data, preprocessing, feature extraction, learning & classification and output data. The system preprocesses the input data from Medline literature and BB'16, and then extracts the features including word representation feature, common feature and Brown clusters feature, based on SVM classifier to learn and classify.

In the GE4 task, the system has three main components: trigger detection, edge detection and post-processing. During the trigger detection, we propose a two-stage method, which divides trigger detection into recognition stage and classification stage. During the recognition stage, we just discern the words which are trigger words, selecting the features that are more suitable for recognition; in the classification stage, we classify the triggers which are identified already, selecting the features that are more helpful to classification. In the edge detection, a muti-class PA algorithm is used, finally the events are obtained by post-processing of TEES.

	Precision	Recall	F-Score
Baseline	61.61%	38.35%	47.27%
Ours	59.91%	39.23%	47.42%

Table 1: Performance comparison on the test set

2 Experimental Results

In the BB3 task, the system achieves an F-score of 56.38% on the development set, which is 4.38 percentage points higher than the baseline. On the test set, it achieves an F-score of 47.4% on the BB3 event task, the result is shown as table 1.

In the GE4 task, the performance of our system is evaluated on the test dataset of the BioNLP'16 with online evaluation. The results related to event extraction are listed on Table 2 and Table 3.

Relations	Recall	Precision	F-Score
ThemeOf	0.51	0.50	0.51
CauseOf	0.22	0.55	0.32

Table 2: The result of relations

Denotations	Recall	Precision	F-score
Gene-expression	0.85	0.88	0.87
Binding	0.68	0.72	0.70
Localization	0.51	0.84	0.63
Phosphorylation	0.86	0.85	0.86
Potein_catabolism	0.85	0.69	0.76
ALL	0.83	0.92	0.87

Table 3: The result of denotations

3 Conclusion

In the BB3 task, our system applies distributed word representation and Brown clusters representation methods, and obtains better performance than baseline, achieving the 4th place. In the GE4 task, we adopt a two-stage method for trigger detection, which effectively avoids the situation that excessive negative samples are classified into positive samples, and the performance of the system is improved. In addition, we select different features in each stage.

Acknowledgment

This work is supported by grant from the National Natural Science Foundation of China (no. 61173101, 61173100).[1]

---

[1]*Corresponding author

Proceedings of the 4th BioNLP Shared Task Workshop, page 101,
Berlin, Germany, August 13, 2016. ©2016 Association for Computational Linguistics

Author Index

Association for Computational Linguistics
209 N. Eighth Street
Stroudsburg, Pennsylvania 18360

ISBN 978-1-5108-2765-3